A Field Guide to Gold, Gemstone and
Mineral Sites of British Columbia

Volume I: Vancouver Island

A Field Guide to Gold, Gemstone and Mineral Sites of British Columbia

Volume I: Vancouver Island

Rick Hudson

ORCA BOOK PUBLISHERS

Canadian Cataloguing in Publication Data
Hudson, Richard (Richard Dennis), 1947–
    Gold, gemstone and mineral sites of British Columbia

    Includes index.
    Contents: Vol. I, Vancouver Island
    ISBN 1-55143-057-6
1. Mines and mineral resources – British Columbia. 2. Prospecting – British Columbia. I. Title
TN27.B7H82 1996   553 4'09711   C96-910287-9

Library of Congress Catalog Card Number: 97-65295

Cover design by Christine Toller
Printed and bound in Canada

Orca Book Publishers
PO Box 5626, Station B
Victoria, BC Canada
V8R 6S4

Orca Book Publishers
PO Box 468
Custer, WA USA
98240-0468

99 98 97   5 4 3 2 1

Front Cover Photo:
Panning for gold at the now-derelict village of Leechtown, near Sooke, site of Vancouver Island's first gold rush in 1864. (Sites 21-24, 26, 28)

Back Cover Photos:
(*Top*) Entrance to the past producer Blue Grouse Mine shows copper sulphides and carbonates, Lake Cowichan. The mine dump is full of pyrite, chalcophyrite, cuprite, garnets and malachite. (Site 36)
(*Centre*) Called variously flowerstone, chinese writing stone, or chyrsanthenum stone, this porphery of dark gabbro with light feldspar crystal "flowers" is found near Chemainus. (Site 43)
(*Bottom*) Several hidden seams of blood-red jasper and quartz cut through the Chemainus River Valley. Large amounts of "float" can be found in the creek, or their sources can be tracked down. (Sites 42-44, 46)

There are many beautiful things in this world: a butterfly's wing, autumn leaves on a stream, a glowing sunset. Alas, all of them will pass, leaving nothing behind. That's probably why, as humans, we value gems and minerals so much; they are breathtakingly perfect, yet almost immune to time. Those tiny shards from the earth are both enchanting and eternal.

To Max Seward,
geologist, philosopher, teacher, friend
who lit the fire in many hearts
and produced so many extraordinary earth scientists.

ACKNOWLEDGMENTS

This book draws from many people's work. In the 1970s, Ann Sabina, Howard Pearsons, Stan and Chris Leaming, Bill and Julie Hutchinson, and Ron Purvis published books about rocks and minerals in BC. They made my task much easier to follow.

Within the Victoria Lapidary & Mineral Society, many people helped, and in many ways. Credit for starting this book rests squarely with Lorena Taylor (a retired teacher) who told me to stop talking about the damn thing and start doing something about it. Magdalene Magon and Gary McCue provided enthusiasm and specimens, Gilles Lebrun shared field trips and introduced me to the magical world of micro-mounts. For over three years Tom Vaulkhard (publisher of the Hutchinsons' book) asked me every meeting when it would be finished!

A special thanks to Simon Deane, my principal field partner, for the many bumpy trails we covered in Thunder Paw (his 4x4), and Grizz (my 4x4) and the many skunked weekends spent searching wrong valleys, slopes or peaks. But we found some interesting stuff, nevertheless! And to my son, Peter, who shared overnight trips in varying weathers, and encouraged me to go the extra metre.

Thanks also to John McDonald, Jim Hawkins and the staff at AXYS Corporation of Sidney, BC, whose excellent QUIKMAP program allowed me to draw the maps shown on pages 113 to 120.

Thanks to Dr. Tark Hamilton and Dr. Chris Yorath at the Pacific Geoscience Centre, for discussions geological. Thanks to the staff at BC's Geological Survey Branch offices in Victoria; in particular, Larry Jones, Dorthe Jakobsen and the MINFILE group, without whom this project would have been impossible. Thanks to Dr. George Simandl for field trips, advice and ideas; and particular thanks to Dr. Nick Massey for long discussions, help with literature searches, for giving freely of his field expertise, and for proofreading the final manuscript. Thanks also to Sharon Ferris, Head Librarian at the Jack Davis Geological Library, who dug up obscure references and chased ancient reports.

Finally, thanks to Max Seward, to whom this book is fondly dedicated, who started my love of rocks (late in life). Although now in his mid-eighties and unable to go on many field trips, his enthusiasm every Monday over our weekend discoveries was a tonic. His knowledge, built up over sixty years of geologizing, settled uncertainties, exploded theories, and (occasionally) revealed a diamond in the rough. Well, not exactly a diamond. Every rock hound needs someone like Max. If you can't find one, join a lapidary club (see Chapter 10), where they abound.

BRITISH
COLUMBIA

I am pleased to introduce this field guide to gold, gemstone & mineral sites of Vancouver Island, not only because this is a book that both the visiting geologist and amateur rock hound can enjoy, but because of its timeliness. It has been a quarter of a century since the last such guide was published. In the intervening time, our understanding of the geology of Vancouver Island has changed dramatically and a number of exciting new fossil and mineral sites have been discovered. This book includes some of these interesting new locations as well as numerous "old classics."

The history of mining in British Columbia is peppered with accounts of dedicated amateurs, prospectors and dreamers who, through their prospecting finds, have contributed to the growth of the industry. Today we can be proud of British Columbia's world-class expertise in finding, developing and financing mines; expertise that is now being applied around the world. Many of these people began their careers working on what, at the time, seemed to be insignificant prospects but which eventually developed into some of BC's major mines.

With new access to the back country, new ideas and a little luck there are still new finds to be made.

Happy prospecting,

Ron Smyth
Director
British Columbia Geological Survey

Ministry of
Employment
and Investment

Mailing Address:
PO Box 9320 Stn Prov Govt
Victoria BC V8W 9N3

Location:
5 - 1810 Blanshard Street
Victoria, B.C.

TABLE OF CONTENTS

INTRODUCTION

The countries of North America have busied themselves for so many years with digging vast quantities of coal, leveling mountains of iron ore, and tapping subterranean lakes of oil, that their considerable gemstone resources have often been ignored.

– John Sinkankas, "Gemstones of North America", 1959

British Columbia is a great place to find interesting and unusual minerals, gems, semi-precious stones, fossils and, of course, gold. In fact, it's the best province for rockhounding in western Canada. The Earth's crust has been pushed and pulled, folded and eroded, squeezed and metamorphosed, so that today there is a wonderful array of geological conditions, offering a wide range of rockhounding opportunities. And it's right here.

In the past forty years, interest in amateur rock collecting has grown by leaps and bounds, and throughout the province there are numerous clubs and associations where folks with a common interest in minerals, gold, crystals, fossils, artifacts and much, much more can get together and yack about the past and plan for the future, while at the same time educating each other.

Minerals are the basis of our civilization. To exist the way we do today requires a bewildering range of materials that are derived, somewhere back down the line, from the earth. Cement, sand, blacktop, aluminum, steel, plastics, glass, silicon, oil, coal and uranium all come out of that thin shell on the planet we call "the surface." Most of the world's troubles can be attributed to the unequal distribution of those resources, and much of our future uncertainty lies in using the remaining wealth intelligently.

Think globally, act locally. If ever there was a time to live up to this motto, it's now, managing the rocks and minerals of the earth. First, be aware of the scale of today's production; next, understand how your own little patch fits into the scheme of things.

This guide, I hope, addresses that second challenge. Welcome to rockhounding in British Columbia!

CHAPTER 1: GEOLOGY

1.1 ALL ABOUT ROCKS AND MINERALS

What are rocks? They are combinations of minerals that in turn are made up of compounds. Compounds are formed from the basic elements. Everything around us, and of us, is made up of just 90 elements (sulphur, gold and iron are typical elements). Together, these elements form a bewildering number of compounds. Fortunately, in the field of mineralogy, there are only about 3,000 different minerals, and we need to know only about 50 to 150 to sound like a real rock hound!

These minerals are found on that thin layer of the earth's outer hard surface known as the crust. The continents are islands, composed of lighter rocks, floating on an inner mass of heavier ones, far below the surface and still at very high temperatures.

The most common elements by weight of the earth's crust are believed to be oxygen at 53%, silicon at 26%, then aluminum 7.5%, iron 4.2%, calcium 3.3%, potassium 3%, hydrogen 1%, carbon 0.4%, and so on down the list. It's hardly surprising, then, that *silica* (silicon dioxide or SiO_2, which is what glass, sand and quartz are all made from), is one of the most common compounds in the crust.

A given rock may be formed from one or several minerals, and its properties will depend on its components. It may be plastic like clay, malleable like gold, powdery like sand, or liquid like mercury.

The study of rocks includes many different fields. The *geologist* identifies the rock formations and determines how and when they were formed. *Petrologists* identify and classify the rocks themselves; the *mineralogist* identifies and studies the minerals that make up the rocks. There are lots of other *-ologists* dealing in other aspects of geology, but we are most interested in these three subjects.

It is well worth learning as much as you can about the science of rocks, even if you believe you will never go beyond identifying the occasional rock found in a river bed. The fact is, just as you don't need to understand how an engine works to drive a car, still, if you do, it will make you a better driver. So too with rockhounding; a basic understanding of the processes and minerals makes you more likely to find 'the big one' that we all know is out there, with our name written all over it! And remember, only a century ago when the first oil prospectors were looking in the United States, they used to throw a hat in the air, and drill where it landed! We've come quite a way in a hundred years!

Learning about rocks can take place at many levels. There's a lot to be said for joining a local lapidary or rockhounding club (see Chapter 10: Addresses). They organize field trips and lots of hands-on learning. For those wanting something a bit more thorough, colleges, institutes and universities offer night classes for more rigorous training.

1.2 THE SHAPE OF BC (MORPHOLOGY)

The earth's surface is constantly being changed, either catastrophically (like Mount St. Helens), or gradually, in the form of erosion. The main agents in erosion are water, ice, wind and chemical breakdown. In the recent past in the province of BC, glacial ice has played a major role in shaping the land and depositing sediments. Today, rivers are the major erosion force.

Bedrock also affects the shape of the land. Whether it's hard or soft, cracked or massive, makes a big difference. Climate has an affect too — in a desert area, limestone will last much longer than in a high rainfall zone, where it slowly dissolves.

There are four major physiographic divisions in the province. Reading from left to right on the map, they are the Western, Central, Eastern, and Alberta Plains divisions.

The Western System is made up of two ranges: the outer mountains of the Queen Charlotte Islands and Vancouver Island, and the inner Coastal Ranges that back onto the City of Vancouver, and run up the coast to the Alaskan Panhandle. The two ranges are separated by the Coastal Trough (Georgia Strait up to Hecate Strait). This is an active area; in the past 10 million years volcanoes have been formed along the margin. All sorts of igneous minerals are there for the finding.

The Central System is a complex mixture of plateau and mountain areas that formed over the last 200 million years, and is the result of recurring glacial ice, as well as volcanic activity. For example, the basalts south of Ashcroft are 200 million years old, while those at Spences Bridge are only half that age. By and large, the mountainous areas of the Central System are underlain with old rocks, including igneous, metamorphic and Pre-Cambrian sedimentary. The plateaux, on the other hand, are generally younger. This is a great mineral hunting area.

The Eastern System is quite different; it is made up of the Rocky and Mackenzie Mountains, and is bounded on the west by the Rocky Mountain Trench, home to towns like Kimberley and Golden. Granites are uncommon; instead, the records show massive piling up of slabs of rock that were pushed and slid from the west, mostly between 100 and 40 million years ago. This created thick piles of sedimentary rocks, such as Mount Rundle and Cascade Mountain at Banff. In some places, old rocks have been pushed up over younger ones. In most, there is a tendency for the southwest slopes to be gentle, the northeast to be steep or even cliff-like, resulting in grand scenery. This is world-class fossil country.

The Alberta Plains is a sedimentary basin that intrudes into the northeastern corner of the province, on the east side of the Rocky Mountain ranges. Drained by the Peace River, sediments here are thousands of metres thick, and are known best for their oil and gas production.

1.3 THE ROCKS OF BC (PETROLOGY)

A noted university professor of geology used to begin his introductory lecture with the immortal words, "There are only two kinds of rocks — ones you can make money out of, and ones you can't."

This is rather an over-simplification! Instead, you could say there are three kinds of rocks: *igneous, sedimentary* and *metamorphic*. Let's explain:

1.3.1 Igneous rocks

Igneous rocks are those that have come from molten matter deep inside the earth. They may be *intrusive*, which means they never reached the surface, but squeeze into cracks on the way up (as *dykes* or *sills*), and are later exposed due to weathering or uplifting. Granite is a typical intrusive rock, being coarse-grained. It cooled slowly, deep in the earth, allowing feldspar, quartz and usually mica and hornblende to crystallize in the rock matrix. Most of the Canadian Shield in Ontario and Quebec was formed this way. Closer to home, the mountains in the Bugaboo Range (to the west of Windermere) are granites, as is the Squamish Chief, the huge cliff that overhangs the highway near Squamish at the head of Howe Sound. Other important intrusives are *pegmatites*, which frequently house the best gemstones, *syenite* which is a granite without any quartz, *diorite, gabbro* and *peridotite*, the latter often containing important commercial minerals. Mountaineers often prefer intrusives because they offer hard, well-cracked rock faces to climb.

Igneous rocks can also be *extrusive*, which means they reached the surface as lava or magma flows and cooled quickly to form fine-grained material. Hawaii and Iceland are on-going examples of this process. The lava flows below Black Tusk Mountain and Garibaldi Peak, above the Squamish-Whistler highway are like this, where you would expect to find *obsidian, basalt*, and *rhyolite*. Rhyolite has pretty much the same chemistry as granite but, by cooling more quickly, does not allow crystals to form and grow. When rhyolite is blown by hot gases into a sponge texture, it is called *pumice*, which is a volcanic froth. Pumice has so many air pockets, it floats on water!

1.3.2 Sedimentary rocks

Sedimentary rocks are varied, and are usually produced by water and sometimes wind, depositing them in layers. Sedimentary rocks are therefore formed in *strata* or beds, and are initially flat. Some are formed from pre-existing rocks by erosion, and can be fine-grained like the sandstone beaches found along the shores of the Gulf and Georgia Strait islands, or coarse grained conglomerates, which look like piles of river pebbles glued together with concrete. The Rocky Mountains along the BC-Alberta border, are fine examples of sedimentary processes, with their obvious "layer cake" look, often tilted due to recent thrusting.

Some sedimentary rocks are formed by chemical precipitation (e.g. some types of gypsum, chert and jaspers), while others are formed by compacting the shells of marine animals. A lot of chalk and limestone cliffs (including the famous White Cliffs of Dover in England) are actually countless marine organisms piled and squished down into layers hundreds of metres thick.

1.3.3 Metamorphic rocks

The third type is *metamorphic rock*, and basically refers to any rock that has been changed after being formed. Thus, both igneous and sedimentary rocks can be *metamorphosed* into something new. The change can be very slight (slate is simply shale that has been cooked in an oven for a few millennia), or it can be so dramatic you can't tell what the original material was (sometimes know as *fubar*ite = 'fudged up beyond all recognition')! Pressure, chemical penetration, and heat are the great metamorphic forces. Marble is metamorphosed limestone, and quartzites were usually sandstones once. *Schists* and *gneisses* (pronounced 'shists and nices') are often heavily changed shales and granites, after other material has invaded or permeated them. Geologists spend years arguing about what happened, when, and to whom.

1.4 A SHORT GEOLOGICAL HISTORY OF VANCOUVER ISLAND

If you are visiting from the mainland, the first thing you'll notice about the rocks on Vancouver Island is that they are very different. No soaring granites (as at Squamish), no cindery volcanoes (as at Garbaldi Peak and Mt. Baker). The island is noticeably different, because it belongs to a completely separate terrane called Wrangellia, which includes the Queen Charlotte Islands, and part of southwest Alaska.

It all started about 380 million years ago in the Devonian era, somewhere out in the Pacific, possibly close to present-day Hawaii. Like those tropical islands, Wrangellia began as a series of undersea volcanic eruptions. Unlike Hawaii, not many reached the surface. Huge amounts of ash (green tuff) spread across the sea floor. Geologists estimate this went on for about 20 million years, after which everything cooled down, leaving the rocks of the Sicker Group, which today are exposed on Salt Spring Island, the Cowichan Valley (including Mt. Sicker north of Duncan, which gave its name to the structure), and the Myra Falls/Lynx mine in Strathcona Provincial Park.

From then until the start of the Triassic era, 245 million years ago, the shallow waters of the newly formed sub-sea islands eroded and weathered, plants and marine animals lived and died, and layers of limestone built up. Today these are visible around Victoria (Sites 2, 6-8), on the west coast at Sites 60 and 63, at Mt. Mark (Site 67), on Buttle Lake (Site 75) and others.

Curiously, there are no rocks on the island from around the start of the Triassic era (245-235 million years ago). Perhaps they all weathered away. However, about 230 million years ago the crust appears to have split apart, and huge masses of fiery lava oozed out, cooling quickly into black basalt (pillow lavas) in what is called the Karmutsen Formation. This is the thickest layer found on Vancouver Island. As more and more lava rose and spread outwards, the entire region was covered in such thick layers that after 5 million years, when it all slowed down, the topmost layers were close to the surface. Perhaps the best example of this is Mt. Arrowsmith, a high, domed cluster of peaks between Parksville and Port Alberni.

As the Triassic came to an end and the begining of the Jurassic saw the arrival of the first dinosaurs, mammals and birds (210 million years ago), more lava flows pushed Wrangellia above sea level, forming the Bonanza Group. At the same time, other magma failed to reach the earth's surface, but in coming close, the heat cooked (metamorphosed) the deeper rocks in the Sicker Group, turning some to gneiss. Evidence of this can be found on Mt. Tolmie in Victoria. Limestone also metamorphosed to marble, as at Port Renfrew.

Up until the late Jurassic, the forces working on what would come to be known as Vancouver Island were mostly eruption and erosion. But something else had been taking place, quietly in the background, for almost 300 million years ... plate tectonics. While volcanoes were spreading thick layers of rock across the sea floor, the sea floor itself (and the volcanoes) were moving slowly and steadily northeast, until about 100 million years ago they ran into the North American Plate, which was itself heading west.

Powerful forces began to twist and shove and lift and squeeze the Wrangellia Terrane. Fault lines developed, such as the ones along the Cowichan Valley, and up Buttle Lake, and mountains rose along the island's centre. Newly eroded material washed into the closing gap of Georgia Strait, forming a series of shallow basins, part of the Nanaimo Group, where coal would later be discovered.

With the Wrangellia Terrane halted by the North American continent, another plate (the Crescent Terrane) came gliding along and slammed into the back of Wrangellia, about 42 million years ago (just like cars on a foggy highway). These new rocks can be seen from Metchosin to Loss Creek, and are known as the Metchosin Igneous Complex.

Since then, most of the changes to the island have been small — an eruption here, a magma flow there. Nothing to write a book about. The final chapter in this continuing saga is the arrival of at least three ice ages, which dramatically changed the surface of Wrangellia/Crescent, leaving it much as we know it today. The last ice sheets retreated only 12,000 years ago, which in geological time, is almost yesterday.

1.5 WHERE TO LOOK WHEN ROCKHOUNDING

There's an old rock hound saying that goes: *Don't look for minerals, but look where minerals are.* The fact is, if you know where the right conditions exist for a particular mineral, chances are you'll find it. On the other hand, if you just sort of head out prospecting, all you'll get are bug bites and strong legs.

Road and railway cuttings are a good source of material, being close to your vehicle, plus they are often active (mini-slides), revealing fresh faces to examine. And, of course, the good old Department of Highways is frequently rebuilding roads and new ones are always under construction, so that major new surfaces are exposed, all without you lifting a shovel! Fact is, some of the very best sites have been found by highway crews while building new roads or upgrading old ones. So, cultivate your local road engineers — they can be invaluable sources of information.

Cliffs are good sites, as are beaches, lakeshores and river beds. Quarries, mine workings and mine dumps are usually productive places to search for minerals, but be careful! **Shafts and tunnels in old mines are frequently unsafe, and should be avoided.** It's usually safe to work the old parts of a mine dump; dig in from the side, if you can, to avoid having to remove the newer overburden. Examine boulders — there may be something there that was missed. When you find something, note the circumstances, and try finding similar conditions elsewhere.

In the field, know which minerals and rocks are associated with what you're trying to find. For example, garnets are heavy, and found in gravel beds; kunzite is found in pegmatite feldspars; and so on.

1.6 ETHICS

Common courtesy dictates that you ask permission before entering a mine site, quarry or other private property. There may be old adits or drifts that are hidden, but poorly protected. Falling down a mine shaft could ruin your whole day. Your access rights are described in Chapter 4: Staking a Claim.

* If the property is privately owned, it is common to pay some daily fee, or percentage of the take.
* Be cautious about bringing along a dog; bears, porcupines and property owners may not share your admiration for the family pet.
* Respect all wildlife. Before you enter an area, try to figure out what you're likely to meet, and how to react. A few moments' forethought can save you a lot of bactine later.

* Be extra careful with fire.
* Close gates.
* Take out your garbage.
* Don't contaminate the creeks.
* Fill in your holes. Animals, including valuable stock, could fall into them. And besides, you don't want to tell everyone where you've been working.
* Stay off crop and grass land.
* No firearms or blasting unless the owner agrees.
* If the owner allows you to prospect for free (many do), then a gift made from something taken off the property (some tumbled stones, or a pretty agate slice) makes a valued "thank you" the next time you visit.
* Tell someone where you're going and when you're coming home. If you change your mind before leaving the vehicle, leave a note face down on the dashboard, so potential rescue parties have a clue which way you went.
* Be honest. Without honesty, the whole access question gets bogged down, and wonderful collecting sites can be put off limits, sometimes for decades.

CHAPTER 2: A SHORT HISTORY OF BC PROSPECTING

2.1 THE GOLD RUSH STARTED EVERYTHING

The 1858 gold rush in BC followed closely on the heels of the frenzy in California. It's true that Natives had discovered gold in the Queen Charlotte Islands as early as 1850, but production was low and the conditions hostile, and little activity followed. In 1852 gold was found on the Thompson River, but for five years little happened. Then, in 1858, seventy-five American miners arrived and set up operations at Hills Bar, below what is known today as Yale in the Fraser Canyon.

Within a year, the gold rush had swelled from a trickle to a flood. In Victoria, the capital city of Vancouver Island (but not the Mainland), the population jumped from 225 to 450 in a single day, and swelled to 5,000 within a month. Any vessel capable of being patched and rendered semi-seaworthy, set sail from California, crammed to the railings with men dreaming of riches in the gold fields of British Columbia.

2.2 DEFINING A PROVINCE

The Governor, Sir James Douglas, quickly realized that if he did not establish a system to control this influx of southerners, both the gold and much of the province could default to the United States, since the border at that time was still very undefined. As a result, he claimed ownership of the gold fields in the name of the Crown, established a system of miner's licenses, and enforced the law. Anyone venturing into the interior to prospect had to buy a license in Victoria (at a cost of ten shillings per month). Under the terms of this license, a man could stake a claim 12 feet by 12 feet.

By 1860 the prospectors had discovered the Cariboo fields, and word quickly spread of incredible finds. At the same time, the placer deposits along the Fraser and Thompson Rivers were being worked out as men moved huge volumes of gravel and sand with hand tools and little else.

Further gold discoveries followed; in 1862 the Stikine; the Kootenays and Leech River (near Victoria) in 1864; Big Bend in 1866; the Peace-Omineca in 1869; and finally the Cassiar district in 1872.

2.3 A NETWORK OF CIVILIZATION DEVELOPED

As the gold rush progressed, the law, in the form of the North West Mounted Police, locally appointed magistrates, Victoria appointed judges, and Gold Commissioners established a system of records and government that was quite unlike the cheerful lawlessness of the California and Klondike rushes. Behind them, the Royal Engineers followed, establishing roads, bridges, ferries and infrastructure.

Once the surface placers began to peter out, most of the miners left for other pastures, leaving behind a string of new towns and transportation networks which were quickly expanded by ranchers, businessmen and fisheries. Logging and sawmills flourished, followed by housing, hotels, entertainment, and more. In short, in just two decades the non-existent province of British Columbia came into being, was included into the new federation of Canada, gained a network of roads and railway lines, a dynamic population, and a booming economy.

There had been early coal discoveries at Nanaimo on Vancouver Island in the 1850s, and with the gold decline in the 1880s, the importance of coal in the interior grew. Lead and silver, often found in association with gold, were also developed, and copper was discovered in large quantities in the early 1900s.

2.4 MINING BUILT THE PROVINCE

Many towns were founded on mining. For example, Yale, Barkerville and Atlin were the result of gold placer mining; Nelson was silver mining; Rossland was built on copper-gold; Kimberley on the Sullivan lead-zinc mine; Kaslo on lead-silver; Greenwood and Britannia on copper; Endako on molybdenum; Nanaimo, Comox, Fort St. John and Cumberland on coal mining.

In just one short century, mining has seen an amazing shift. Where the old sourdoughs struggled into the interior of the province over terrible trails, enduring scorching summer heat and flies, and arctic winters that killed them in droves, today mining is the forte of large corporations who can raise the huge sums of money needed to move 100,000 tonnes of rock a day to get at the minerals.

2.5 THE LITTLE GUY IS STILL A KEY PLAYER

Yet the prospector still has an important role to play. Despite all the sophisticated technology used in modern exploration, new mineral showings are still being turned up by prospectors; just ordinary people with determination, an independent streak, and a dream. The province is vast, its potential barely tapped. It could happen to you, just as easily as it happened to Billy Barker at Barkerville.

CHAPTER 3: PROSPECTING TECHNIQUES

*Go, my sons, buy stout shoes, climb the mountains, search the valleys,
the deserts, the sea shores, and the deep recesses of the earth. Mark
well the various kinds of minerals, note their properties and their
mode of origin.*

– Petrus Severinus, 1571

Rockhounding can take you into the wilderness, so here are some tips to
make things more comfortable when you're out there.

* Don't prospect alone. Three is the best number — in the event
  of an accident, one can stay with the patient while the other
  goes for help.
* Make sure your vehicle is in good shape; there might not be
  anyone along for a while to find you if you break down. Four
  wheel drive is a lot better than two. And have enough gas!
* Bring food and water; you may want to stay longer than
  planned. Include candles for light and warmth.
* Carry decent maps — the 1:250,000 and the 1:50,000 series are
  useful, and give plenty of detail.
* Get local information from the Gold Commissioner's Offices,
  local rock shops, and rockhounding clubs.
* Don't be too ambitious — hounding takes time, and the more
  time spent driving between sites, the less time to hunt for that
  prize specimen.
* Don't ignore the seasons. Winter in the interior is serious. Even
  summer in the high country can be bitterly cold. Digging hot
  gravels in mid-summer should be done in the cool of the
  morning and evening.

3.1 WHAT TO WEAR

You should wear the same sort of clothes as you would to hunt or hike.
Tough, comfortable shoes or boots are a must; when working near water,

take along a pair of gumboots. If you don't have to worry about getting your feet wet, you'll often go into places that others may have overlooked. Spare socks are appreciated by everyone when you get back to the vehicle.

A canvas or nylon vest full of pockets is handy. They can be bought at most outfitting stores (for fishermen), or ordered from a mining supplier, where they are called geologist's vests.

A lot of prospecting involves bending over. Knee pads are sometimes handy (the skateboard type are great), and a waterproof jacket to protect your horizontal back is a must if you plan to have some comfort. A spare sweater is also handy — gullies, cracks and mine shafts can be cool or downright freezing, even in mid-summer.

3.2 WHAT TOOLS TO CARRY

Take a pair of safety goggles; you can buy them at hardware stores, or a pair of old ski goggles work well. Wear them whenever chipping rock. Work gloves are necessary. Insect repellent is also handy. Some people prefer bug-hats with a wide brim and mesh. Let's face it, when things are bad, nothing seems to work. It makes you wonder how the sourdoughs managed a hundred years ago.

Brawn Tools: a hammer is essential. One with a chisel-shaped back is handy for trimming specimens. A carpenter's hammer is no good — it will chip. A prybar, a shovel, a set of hardened chisels, a smaller trenching tool, a pointed pick, and a spoon are all useful for different jobs. A heavy hammer (8 lb. is about best) can convince hard samples to relax a bit. A small plastic gold pan weighs nothing, yet can provide hours of fun. But remember, the more you have to carry, the less distance you'll be prepared to hike from the road.

Brain Tools: a tough bag with a shoulder strap is indispensable, together with a pocket knife. A leather carpenter's apron is preferred by some. A hardness testing kit (see Chapter 5.1 for materials to use) is handy for identifying specimens by hardness. A magnifying glass (*loupe*) in the X10 to X16 range is useful for checking inclusions and crystallography.

Bring a flashlight, and make sure it can't switch on by accident (put a disk of cardboard between the spring and back battery). A compass and maps are key. I find I use an altimeter a lot, not only to figure out how far up-slope I am, but as a barometer. Overnight, it is a much more reliable gauge of the weather than the radio (if the altitude apparently drops, the weather is improving). Bring newspaper to wrap your specimens (and to read when your partner is still away up the hill). A magnet and streak plate (unglazed bathroom tile) are helpful in identifying an unknown sample.

If you're into chemistry, a dropper bottle containing one part concentrated hydrochloric acid to nineteen parts water (a 5% solution) is handy for testing reactivity. Just make sure that cap doesn't leak, or you'll be looking for new pockets.

A notebook and pencil (you can buy Write in the Rain paper from some stationers), camera and labels are a good idea to keep accurate records. You might also like to know that the average person's index finger measures approximately 2.5 cm (1 in) between knuckle joints, which means that you can measure the width of a mineral vein, or the length of a crysotile fibre with considerable accuracy, simply by bringing a hand along on field trips.

A recent arrival on the scene is the Global Positional System (GPS) receiver. About the size of a calculator, it receives radio signals from up to twelve orbiting satellites, and calculates its position in latitude, longitude and altitude with an accuracy of between 30 metres and 300 metres. While this may seem an unnecessary refinement when you are working in open, flat country, in the thick forests and box canyons of BC you can walk within 50 metres of a site and never even see it. GPS receivers (now under $300 and the prices are still dropping), when used with common sense, can save you a lot of wasted time.

And, of course, bring this book!

3.3 CARING FOR YOUR SPECIMENS

There's always a temptation to take too much, with the result you end up with a basement full of stuff you'll never really use. That's not fair on the next guy to try his luck, and it often irritates the landowner when he sees huge chunks of his property disappearing into the back of a pickup — sort of makes him wonder what he's missing. Trim your specimens to a practical size. Wrap them individually in newspaper, and label them with place, time, and other relevant information. If you don't, it's funny how they all look the same when you get back to the road and start gloating! Pack them carefully, to avoid later disappointment, especially if the hike out is rough.

Once home, wash the specimens in water. Don't use stiff brushes on soft materials. Label each clearly. You can paint a small white patch on an inconspicuous surface, using typist's white-out liquid, and then number with an indelible felt-tip pen. Note all the details in your field book.

As your collection grows, you may want to arrange it in some sort of system (by area, or date, or colour, or material), so loose labels may get misplaced, and numbering sequences interrupted. Many serious collectors use Dana's System of Mineralogy which is a system of classification of minerals based on their chemical properties. Developed by James Dwight Dana, a Yale professor, it was first published in 1837 and significantly revised in 1854.

3.4 SAFETY TIPS

Old mine sites are hazardous places; coal shafts fill with odorless methane gas.

* When entering old workings, LOOK UP! Have a buddy within earshot. Don't hammer or walk upslope of your friend.
* Wear safety glasses, and check that your hammer's head is secure.
* Before visiting for the first time, talk with someone who knows the area. Features change, new roads make easier access.
* Carry a first aid kit. And check it's complete BEFORE you start.
* Respect the weather. It's bigger than you.

CHAPTER 4: ACCESS RIGHTS

4.1 ACCESS RIGHTS

First, a bit of legalese to explain what rights exist for property owner and prospector:

Most land owners in BC hold clear title to their property, but their Certificate of Title seldom includes what lies beneath the surface. Surface and subsurface rights are separate and distinct. Unless otherwise excluded by the property title, the land owner's rights usually only extend to soil, sand and gravel. Under normal circumstances, the Crown reserves the rights to the subsurface, which may include coal, petroleum, natural gas, and minerals (both base and precious). The rights to these substances may be granted by the Crown separately under different legislation. The land owner's rights are defined in the **Land Act**. Any person disregarding those rights is subject to the **Trespass Act**. A person may be entitled to enter private lands, however, in order to acquire subsurface rights, under the **Mineral Tenure Act**.

Many of the sites listed in this guide are on unoccupied Crown Land, so you do not need any paperwork just to have a look around. Anyone can hand pan or prospect on most Crown Lands, with the following exceptions:

1) in parks,
2) on a valid Placer Claim or Placer Lease (see section 4.3 for more details),
3) on an Indian Reserve,
4) on land occupied by a building,
5) up to 75 metres around a dwelling house,
6) orchards and cultivated lands.

Recreational hand panning and prospecting are allowed without having to be a Free Miner (see section 4.2 below for definition of a Free Miner).

Hand panning equipment is restricted to a hand shovel and a hand pan. The key word here is *hand*. You can't use mechanical equipment for recreational purposes. In particular, portable suction dredges are strictly regulated in BC; you should check with your nearest Gold Commissioner before taking one into the field.

4.2 GETTING SERIOUS

Even if you plan to just hunt for interesting "stuff," you should know what your rights are when crossing private and Crown Lands. This section is for that magic day when you find "the big one," and reckon there's enough neat "stuff" at that site that you'd like exclusive rights to commercialize it.

The first step is to contact your local Gold Commissioner to get current mineral titles information on the area where you plan to work. The Gold Commissioner will provide maps, tags, staking regulations, forms, other information, and will, of course, take your fee too.

You may need to become a *Free Miner*, which sounds great, doesn't it? ("You can't arrest me, officer, I'm a Free Miner.") To qualify, you must be a permanent resident, and be 18 years or older (over 65 is free). A Free Miner can also be a corporation, or a combination of individual and corporation. Any Gold Commissioner can issue a *Free Mining Certificate*. The fee in 1996 is $25 for an individual, $500 for a corporation. The fee is subject to increase, like everything else in government. Shortly, there may be a written examination to prove you know and understand the contents of the Free Miner's Handbook.

A Free Miner has the right to enter mineral lands for the purpose of mineral and placer mineral exploration. You can locate a claim on privately owned land. However, a Free Miner's right of entry onto private land does not extend to land under cultivation, orchard and pasture land, land occupied by a building or the land immediately around it. It is recommended that you contact the land owner before any staking activity. You might not get unconditional approval, but you will have made the first legitimate step, which is important. The law has, over the years, defined pretty clearly (from a lawyer's point of view) what you may and may not do. In theory, you don't have to get permission from the land owner unless you want to explore, develop or produce using mechanical means. The land owner can appeal to the Gold Commissioner if you want to disturb the surface by mechanical means, and you may be told to post a bond. A Free Miner is responsible for any loss or damage that may be caused by his use or entry of private land.

A Free Miner may take samples for assay without the complexity of getting formal approval for an extraction permit from the Gold Commissioner's office. It may come as a surprise to you to learn that a "sample" may be up to 10,000 tonnes! Consult your local Commissioner's office for more information.

4.3 THE MINERAL TENURE ACT

This act was introduced in 1988, and replaces the two previous acts governing the staking of claims: the Mineral Act, and the Mining (Placer) Act. It is administered by the Mineral Titles Branch of the Ministry of Employment and Investment (MEI used to be Energy, Mines & Petroleum Resources — see Chapter 10: Addresses for further details). One of the main changes is the administration of placer claims. A new title, the *placer claim*, was introduced, that can be located in a greatly expanded area of the province.

There are now five basic types of mineral title:

1) mineral claims
2) mining lease
3) placer claims
4) placer leases
5) lease of placer minerals

Both 1 and 3 are intended to permit exploration and small-scale development. Mining leases (2) and placer leases (4) are designed for large-scale production by the big guys.

4.4 HOW TO STAKE A CLAIM

The law changes; interpretations of the law change too. For the most up-to-date information, you are strongly advised to contact the MEI in Victoria, or any of the local Gold Commissioners (Chapter 10: Addresses) for the latest changes.

Follow the procedures correctly. At best, a slight omission can sometimes mean a complicated court case or possible forfeiture of a claim.

4.5 WHAT IS A MINERAL?

Under the Act, a *"mineral"* refers to ore bodies of metal and other natural substances that can be mined (including *mineral substances* such as limestone, dolomite, marble, shale, clay, volcanic ash and diatomaceous earth), that occur either in the place in which they were originally formed or in talus rock adjacent to the place where they were formed.

In addition, materials from mine tailings, dumps and previously mined deposits are also included. On the other hand, coal, petroleum, natural gas, earth, soil, peat, marl, sand and gravel are not included under the Act.

Mineral titles give rights to both mineral substances and other non-placer minerals, provided title to the mineral substances is held by the Crown. Mineral substances held by land owners prior to the 1988 Act cannot be acquired by staking.

Placer minerals are ores of metal and other natural substances that occur either loose or in fragmented rock. They are found in loose earth, gravel and sand, having been transported from the original bedrock source by natural means such as glacial or water action. They also include materials from placer mine tailings and previously mined deposits of placer minerals.

4.6 GETTING STARTED

To stake a claim, you must purchase the appropriate metal tags (*mineral claim* or *placer claim*), available at any Gold Commissioner's or Sub-Recorder's office. These tags have a serial number and must be for the type of claim being staked. The types of claim are:

* four post mineral claims
* two post mineral claims
* fractional mineral claim
* placer claims

Naturally, don't buy tags until you've found a site that you think is worth tagging. Many rock hounds spend years (sometimes their whole lives) hunting and digging without ever bothering to settle down on a particular site for "the big dig."

4.7 WHAT IS A POST?

Posts are used to establish the exact position of claims. As a result, they have to have certain minimum specifications. For a start, there are two kinds: *legal posts*, and *identification posts*. To set up a four post mineral claim, you must use both legal and identification posts. The other types only need legal posts.

A post can be a piece of sound timber, a stump or tree cut to the same specifications, or a cairn of stones. It must be at least 1 metre (40 in) above the ground, and it must be squared and faced on four sides for at least the top 25 cm (10 in). The top must be at least 8.9 cm x 8.9 cm (3.5 in x 3.5 in) square for a legal post, or 3.8 cm x 8.9 cm (1.5 in x 3.5 in) for an identification post. They must be securely placed in the ground. If they aren't secure, or of the correct height and shape, your claim could be invalidated.

You cannot stake a claim between midnight and 7 am, and there are certain areas where you cannot stake at all: Indian Reserves, parks, land occupied by a building, land surrounding a house, land under cultivation, reserves set aside by the Mineral Tenure Act (Minerals Reserves). If you stake a claim that partly overlies any of these, the claim is valid, but you only acquire what is outside the reserved areas. **You must not place posts or blaze a boundary or location lines within any Mineral Reserve**.

Before cutting or blazing trees on private property, contact the owner or occupant of the surface.

4.8 STAKING A MINERAL CLAIM

Staking a *four post claim* is a major undertaking, and you are advised to consult with your local Gold Commissioner for exact details. A four post claim is designed for someone, such as a mining corporation, interested in large areas. It's sufficient to record here that you can enclose from 25 hectares to 500 hectares (about 60 acres to 1,250 acres).

Under the Act, there is no limit to the number of *two post claims* that a

Free Miner can stake in a year. A Free Miner can also stake two post claims as an agent for other Free Miners.

Figure 4.1 shows an example of a two post claim. The maximum size is 500 metres x 500 metres, or 25 hectares (about 62 acres). Each claim is designated as one *unit*.

The position of a two post claim is defined by placing one post at each end of a *location line* as shown in Figure 4.1. The length of the claim is measured along this line, and cannot be longer than 500 metres, nor can it change direction. The width of the claim is measured at right angles to the location line, and cannot be longer than 500 metres to left or right or both left and right.

4.8.1   Procedure

Figure 4.2 shows the procedure for a two post claim.

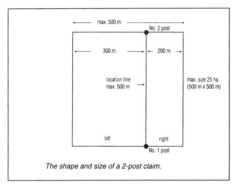

The shape and size of a 2-post claim.

1) Place a legal post at one end of the location line. This becomes No.1 post.

2) Fix to this post a metal tag issued for two post claims, and embossed with the words "INITIAL POST (NO.1)" on the side facing the location line.

3) Mark the Initial Post metal tag with the relevant information.

4) Mark the location line by blazing trees on the sides facing the line, and by cutting the underbrush. On private property, contact the owner before cutting or blazing, obviously!

5) Place another legal post at the other end of the location line, not more than 500 metres from the No.1 post. This becomes No.2 post.

6) Fix "FINAL POST (NO.2)" metal tag to this post, facing the location line.

7) Mark the tag with the relevant information.

If you use a cairn of stones, the tags must be placed securely inside the cairn.

Sometimes, it is impossible to place a final post, because of water, ice, or a cliff. In such a case, use a "*witness post*" for the No. 2 post. (You can't use a witness post for your No. 1.) In such a case, follow the instructions in steps 1 to 4, then:

5) Place a legal post as close to where you want No.2 post to be. This becomes the witness post.

6) Mark this post on the side facing away from the No.1 post with the words "NO.2 WITNESS POST", together with the bearing and distance to where the No.2 post should be.

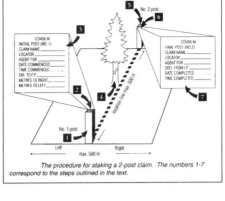

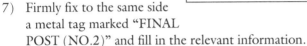

The procedure for staking a 2-post claim. The numbers 1-7 correspond to the steps outlined in the text.

7) Firmly fix to the same side a metal tag marked "FINAL POST (NO.2)" and fill in the relevant information.

Fractional claims occur when much of the surrounding territory has already been staked, and there are odd slices or slivers available. In such a case, obviously you can't always have a 500 metres location line, and a rectangular shape. Fractional claims can only be located over areas of 25 hectares (62 acres) or less. In such cases, the boundaries of the pre-existing claims become the fractional claim's boundaries. The procedure is the same as for a two post claim, except the letters "FR" must be added after the name of the claim on the tags. Witness posts cannot be used.

4.9 STAKING A PLACER CLAIM

Placer claims can only be located in areas designated as Placer Claim Lands, or Placer Lease Lands. Check the Placer Titles Reference Maps at the Gold Commissioner's office closest to you. A Free Miner can stake any number of claims per year.

The main difference between a Placer Claim and a Mineral Claim (section 4.8) is that a Placer Claim may have a location line 1000 metres long (as compared to one of 500 metres), but with the same 500 metres width, making a total area of 50 hectares maximum (125 acres). You will need placer tags, not mineral ones, but the placing of posts follows the same steps as for section 4.8.

4.10 AFTER STAKING YOUR CLAIM

Both mineral and placer claims acquire title as soon as the staking is completed, but you must make an application to record your claim(s) within 20 days of completion. Apply at the Gold Commissioner's or Sub-Recorder's office in which the claim is situated. You can mail your application, but it must be received within the 20 day period. The application must include:

* a complete application form
* a sketch map
* the prescribed fee

Fax applications are not accepted. If your claim overlies other claims, the claim is still valid, but the bits that overlap are not included in your claim; they remain with the previous holder(s).

4.11 MAINTAINING YOUR CLAIM

A claim is valid for a year. The anniversary date is the day that you did the original staking. To maintain a claim, the holder must, before the anniversary, either perform, or have performed, and record, exploration and development work on the claim, or pay cash in lieu of work, and pay a holding fee.

About two days of effort is all that's currently needed to show you have performed work, but you'll have to keep clear records. Obviously, any bills or other expenses incurred should be filed too. Excess expenses incurred can be carried forward up to ten years for mineral claims, and five years for placer claims.

There's a nice story I heard years ago about maintaining a claim: this guy had a favourite fishing lake way back in the boonies, and over the years he grew so attached to the place that he decided to apply to Crown Lands to buy the lakeshore and build a cottage. No way; the land was not for sale. So he applied to build a summer cottage there. No way; you can't build on Crown Land. So he became a Free Miner, staked the area, built a beautiful cottage on the waterfront, and wrote the cost of the project off against his claim maintenance!

You can also apply to have adjoining mineral titles combined as a *mineral group*, or in the case of placer claims, as a *placer group*. This allows you to spread extra work done on one claim to those adjoining it.

Before you start bringing machinery into an area, remember that the Mines Act states that if mechanical equipment (a pretty loose term) is to be used on a claim and the surface is to be disturbed, then:

*   you must submit a Notice of Work and Reclamation, and
*   a permit must be issued to you

You must also advise the owners of private property before disturbing the surface. You are liable to compensate the owner of the surface for loss or damage caused by your entry, occupation, use of the surface, or use of any right of way. In short, make sure you both know what you're getting into, and make sure it's clearly signed and sealed before a D-9 Cat crawls all over the tundra.

4.12 FORFEITING A CLAIM

If a claim is not maintained as per section 4.11, it is forfeited back to the Crown on the anniversary date, and is available for claiming at 7am on the next day. If a co-owner fails to contribute his portion of maintenance, his rights may be taken by the partner(s) who has met the requirements (Section 30 of the Mineral Tenure Act).

CHAPTER 5: ROCK & MINERAL IDENTIFICATION

Most gemstones are minerals, made up of elements or compounds, occurring naturally. This means they are not manmade or otherwise artificially manufactured; they have a definite chemical composition; often, they have a distinct crystal form. They differ from rocks, in that rocks are made up of minerals in various mixtures. When I explain that concept to kids, I use the analogy of pizza — olives, ham and cheese are the "minerals" and always look the same; the final mixture or "rock" is the pizza itself, and can vary quite considerably while still having the same ingredients.

There are exceptions to the rule above, of course. Amber is formed by plants; pearl is manufactured by oysters, and they're both gems. Granite and unakite are rocks without crystal form (or *habit*), but are often made into jewelry.

In case you've ever wondered why minerals and rocks sometimes have such wacky names, since 1959 the International Mineralogical Associations' Commission on New Minerals and New Names calls the shots. If someone wants to have a name recognized internationally, they must submit a full description of the new species, complete with chemical composition, properties, and a lot more. At the same time, they can propose a name of their choice. Then, after suitable mumbling, the name and species are adopted, or not.

There are two schools of thought when it comes to naming a new mineral: rational and irrational. Rational names are based on some property of the new species; for instance, 'fluorite' fluoresces, 'naphoite' is made of sodium (chemical symbol Na), and phosphate (PhO). Incidentally, the "*ite*" on the end of so many minerals and rocks comes from the Greek word *lithos*, meaning rock or stone.

Irrational names don't tell you anything about the material itself, but usually honour some person or thing who may have contributed to its identification, or was the finder of the first sample. Examples are *bauxite*, named after the region near Baux in France where it was first identified;

wicksite is named after Dr. F.J.Wicks, curator of the Royal Ontario Museum.

To date, about three thousand minerals have been identified. Of these, less than a hundred are gemstone material, and you'll be lucky to see more than thirty in a jewelry store. So, don't despair, it's not as complex as it first seems!

GEMSTONES — A SPECIAL CASE

Gem material must be rare, beautiful, and wear resistant. Fashion also plays an important part in the demand for gemstones. In ancient Greece, yellow stones were much in demand, but in Roman times green was considered the epitome of fashion. Today, clear brilliance is highly prized ("a diamond is forever," etc.). Rarity makes a stone precious, but if it is too rare, not enough people know about it, and there is little demand.

The trouble with most gems and minerals is that in the rough they don't look anything like they do in a jewelry store. Ah, the prizes we've walked over in big boots, and never even suspected. To solve that problem, rock hounds have developed a number of field tests. Some gems can be identified by their colours, but many have weathered or oxidized outer shells and so have no colour at all. Others have lots of colours, which is just as confusing. So identification is attempted through elimination.

Many gemstones are crystals, and so show a distinct arrangement of their smooth faces and the angles formed by those faces. These lattices are the result of the atoms aligning themselves in repetitive patterns. Crystallography is an extensive subject, and not something that can be covered in a field guide like this. In basic theory, crystals grow perfectly, usually over prolonged periods of time. To give you some idea of the rate of crystal growth, consider a surface that grows by 1 mm per day. (That's about the thickness of a fingernail.) To maintain this incredibly rapid rate, over one hundred atomic layers have to align themselves (perfectly) *per second*! It's hardly surprising that crystals are rare, small, and take a long time to form.

In the real world, of course, there are impurities and other factors that result in crystals growing in all sorts of variations. Mineralogists have reduced all crystal shapes to just six systems:

isometric – fluorite, octohedral diamond, trisoctahedral garnet, halite, galena

tetragonal – zircon, rutile, cassiterite

hexagonal – quartz, apatite, calcite, tourmaline, beryl

orthorhombic – barite, topaz, sulphur, staurolite, olivine

monoclinic – orthoclase feldspar, gypsum, epidote, mica

triclinic – rhodonite, plagioclase feldspar

There are exceptions to every rule, of course: opal and obsidian, for example, don't have a crystal structure at all, and are just gemological Jell-O.

5.1 HARDNESS

Mohs' scale of hardness is named after the German mineralogist Friedrich Mohs (1773-1839) who devised a simple standard of comparison, ranging from 1 (soft) to 10 (hard) that determines relative "scratchability." Anything having, for example, a hardness of 6, will scratch anything having a lower number of 5 through 1. Minerals having typical hardnesses are:

1 talc
2 gypsum
3 calcite
4 fluorite
5 apatite
6 orthoclase
7 quartz
8 topaz
9 corundum (ruby, sapphire)
10 diamond

The easiest way to remember the sequence is *"The girls can flirt and other queer things can do."* Gypsum is harder than talc, but not twice as hard. Fluorite is harder than calcite, but is less hard than apatite. If a material scratches everything up to fluorite, but not apatite, then it has a hardness between 4 and 5. For field work, you can buy hardness pencils that come equipped with known stones, or you can make your own. Other useful hardness tools are:

2.5 fingernail
4 penny, common nail
5.5 window glass, penknife blade
6.5 metal file
7 quartz crystal

Remember not to scratch a prize specimen in a visible place! When examining a scratch, make sure you are looking at the scratch, and not some loose powder that has been scraped off. Wash the specimen and see that the scratch scores the real rock.

5.2 COLOUR

Recognizing a rock by its outer colour can be pretty deceiving, because surfaces oxidize and tarnish, fooling the most experienced prospector. You have to look at a clean break to get an honest sample, and even then, there are plenty of variations on the colour theme:

Asterism is the result of an alignment of atoms in the stone so that the reflected light is in the form of a star (four or six points are common). Star rubies and quartz are well known.

Chatoyant is a form of asterism where a single line of light is reflected, forming a cat's eye effect. The most valuable gem is chrysoberyl, although tiger eye is more common.

Colour change is how a material changes colour as the light falls on it from different angles. Labradorite and tiger eye are typical examples.

Colour play is how an opal's fire changes as the stone is turned; the stone's prismatic lenses catch the light and reflect it in various colours.

Fluorescence is the ability of a mineral to glow in the dark when exposed to an invisible light, such as an ultra-violet lamp. An obvious stone is fluorite, named for that very reason.

Iridescence is the play of light produced by tiny cracks and fractures in an otherwise transparent crystal. It's not uncommon to see rainbows in quartz caused this way.

Opalescence is a milky or mother-of-pearl appearance that seems to glow from the inside. Australian fire opal often looks like it's being lit from within by a hot coal.

Phosphorescence is that property that allows a stone to radiate light even after the light source (a bright lamp, for example) has been removed.

5.3 LUSTRE

Lustre depends on the absorption, reflection or refraction of the light by the mineral's surface. Common terms are:

adamantine – brilliant, glittering, like a diamond.

greasy – appears oil smeared.

metallic – smooth and gleaming like a typical metal. Hematite and galena are typical.

resinous – looks a bit like egg-shell porcelain.

silky – looks fibrous, like asbestos.

vitreous – smooth and glossy like glass.

5.4 SPECIFIC GRAVITY, OR DENSITY

Specific gravity (or SG, also called density) is the ratio of the weight of a known volume of mineral, compared to an equal volume of pure water. Water has a density of 1. Most rocks have an SG between 2 and 5, but some minerals are a dead give-away when you pick them up. There's no substitute for a gold nugget — even a pebble pup can recognize it because its SG is 19.3! That's heavy! Other dense minerals are cinnabar (8) and uranite

(9.5). At the other end of the scale, pumice is a volcanic rock containing gas bubbles, and has an SG of 0.9, so it's one of the few rocks that floats!

SG is best done in a lab, although you can rig up a density scale quite easily if you have a triple beam balance at home. It's usually used as a 'clincher' to identify a mineral when all other tests have already been done to reduce the options. Weigh the sample in air (*air weight*). Let's suppose it weighs 327 gm. Now tie a thin thread around it and lower it into a bowl of water so it is completely submerged. Tie the top end of the thread (somehow) to the scale's weigh pan. Suppose the scale now reads 206 gm (*water weight*). Then the SG = the air weight divided by the difference between the air and water weights = 327 / (327 – 206) = 327 / 121 = 2.7

Another way to measure SG is to make up liquids of different densities. Since most rocks of interest are in the range 2 to 5, that does not make it too difficult. Years ago Archimedes realized that an object that just floated in a liquid, had the same density as that liquid. That's why we just float in sea water — we're about the same SG of 1.024. In fresh water (SG = 1) we frequently sink. So, by dropping your unknown specimen into denser and denser SG mixes, you can determine fairly accurately what the density is.

5.5 CLEAVAGE
Cleavage has nothing to do with ladies in cocktail dresses. It's the way some minerals split along planes related to the molecular structure of the mineral, and parallel to possible crystal faces. Cleavage perfection is described in five steps from *poor* (such as bornite), *fair, good, perfect*, and *eminent*, such as in mica, which will flake and flake into thinner and thinner slices. Materials like garnets and galena cleave into beautiful cubes, while calcite cleaves into rhombohedrals. Diamonds have perfect cleavage, meaning a diamond can be split along planes to form the classic octahedral.

5.6 FRACTURE
Any break along non-cleavage planes is called a fracture. There are several characteristic fractures. A *conchoidal break* leaves a clam shell shape of arcs. Obsidian or flint (often used by early peoples worldwide for arrow and spear heads) is typical. An *earthy break* looks ragged, like a broken brick.

5.7 STREAK
Streak refers to the colour left when a material is rubbed across a piece of unglazed tile or porcelain. In metals, especially, the streak may be a clue to the mineral. For example, hematite (a ferric oxide mineral with as much as 70% iron) streaks red, not metal/silver, as you'd expect. As a result, the Greeks called it "bloodstone;" hence the name *haema*(blood)*tite*. Streak only works, obviously, on minerals softer than an unglazed tile (about 4.5).

5.8 MAGNETISM
Only a few minerals are magnetic, but since it's easy to test for, and the results can be both fun to watch and exciting to find, it's mentioned here.

Don't, whatever you do, shove a magnet into the material. If you do, you could be picking chips off that magnet for hours! Instead, place your sample on a piece of paper, and slide the magnet underneath. It'll be obvious if any part of the mineral is magnetic — it'll slide after the magnet but, thanks to the paper, won't stick to it.

Why is it exciting to find magnetic material? Well, the gold rush sourdoughs knew that placer gold was often found in association with black sand, which is powdered magnetite. So, if the black sand you pan is magnetic, you may be getting close to a hot streak!

5.9 ELECTRICAL PROPERTIES

Some minerals, such as amber, sulphur and topaz develop a static charge when rubbed (they make your hair stand on end, if held close to your arm). Other minerals, such as quartz, are *piezoelectric*, meaning if you apply mechanical pressure, they produce an electric charge. Vice versa, if given an electrical charge, they produce a mechanical push. This property is exploited in radio crystals and digital (quartz) watches, but requires special equipment to detect.

5.10 RADIOACTIVITY

Measuring radioactivity is not something the average rock hound is likely to get into. Suffice to say, certain "heavy" elements break down over a period of time (seconds to centuries), giving off energy particles that can be detected using a Geiger counter. Uranium and thorium are probably the best known elements.

Actually, lots of minerals are slightly radioactive. The granite slabs facing city buildings, for example, push out low levels (that are within human tolerance). Some river sands are radioactive too.

CHAPTER 6: ROCK & MINERAL DESCRIPTIONS

Included here are short descriptions of some minerals and rocks. More thorough descriptions and additional information are available from your local bookstore or a public or institutional library.

AGATE

Agate is one of the most popular BC gemstones, and is found throughout the province. It is a quartz with microscopic crystals (*cryptocrystalline quartz*) which gives each piece its unique colouring. It belongs to the *chalcedony* group, a waxy, smooth form of quartz, often found lining cavities, filling cracks, or forming crusts. Agate can be *banded*, usually in concentric circles with the bands wavy or smooth, depending on how it formed; or it can be *brecciated*, where it has other crystals or material included in the matrix. Petrified wood is often an agatized wood.

Agate has a hardness of about 6.5 (same as a steel file), so you can't scratch it with a knife. When held to the light, it looks like heavily glazed glass. In the rough, it is deposited in veins and seams, or may form egg-shaped nodules. When found in place (in the hills) it usually has a hard, nondescript "rind" on its outside that can disguise it well; when found in river beds and along beaches, the outer shell is often stripped off, and some of the colour may show through.

Agate is incredibly varied, with names to describe its forms: *lace, fortification, plume, polka dot, dendritic, moss* and more. In all, the dots, layers or patterns are inclusions. The plume (French for feather) and moss are the result of hot liquids being squeezed into cracks, leaving fern-like shapes as they cool. To be worth collecting, agate must be free of cracks (so no wild hammering on a nodule!). Since it's quite a common material, the quality of the individual piece depends largely on the colour, any interesting patterns, and how it was cut and mounted. Agate slices well into thin, translucent slabs, and can make beautiful matched bookends. In jewelry, agate looks great in cabochons or pendants. Because every piece is unique, many collectors treasure good agates as much as gems like diamonds and sapphires.

ALABASTER

Alabaster isn't technically a gemstone, being really just a massive, fine-grained gypsum, but it is easily worked, having a hardness of 2, and historically has been shaped into vases and other decorative forms. Workshop tools are all you need, and the material's brilliant whiteness makes it an attractive addition to any collector's cabinet.

AMBER

Amber is usually found in BC in association with coal deposits. It is generally believed that amber is a fossil resin, of vegetable origin, that was buried and metamorphosed under special conditions of heat and pressure. Amber often contains inclusions of sticks or insects, which makes it all the more interesting. The colour varies from gold through amber to brown. It's pretty soft at 2.5 on the Mohs scale, so can be worked with ordinary tools.

Amber is enjoying a lot of renewed interest recently, and amber necklaces and other beadwork are popular. The problem is, there are lots of plastics that look like amber, and it's not always easy to recognize the real thing. Two quick tests are: rub the sample on a woolen sleeve and see if it becomes electrically charged enough to attract tiny pieces of Styrofoam (amber is electrostatic); alternatively, with a density of only 1.1 amber will float in very salty water, while most plastics will sink.

AMETHYST

Amethyst is a purple quartz, much prized by BC collectors until huge amounts were shipped in from Ontario and Brazil, so that today amethyst doesn't get the respect it used to. Still, finding amethyst in BC is a challenge, and well worth the effort, since the material can be beautiful both in the rough (as clusters of crystals) and when cut as a February birthstone.

Like quartz, it has a hardness of 7, making it suitable for jewelry. The deep purple (possibly manganese) is considered the best. Amethyst has a long and distinguished history, dating back as a prized stone to Phoenician times. The Romans, for example, believed that amethyst kept the wearer sober, although the literature is not specific as to whether this meant the wearer constantly refused a drink, or was capable of holding it when he accepted!

ANALCITE (ANALCIME)

Forms as trapezohedral or cubic crystals in cavities in volcanic host materials (such as basalt). Ususally white with a greasy appearance. Fragile, but quite hard (H = 5). Rare.

ANDALUSITE

A silica compound, often found in slate as a product of metamorphism, the crystals are thick prisms in turbid gray, brown or opaque. One variation, *chiastolite*, forms a crystal cross. Transparent crystals very rare. SG = 3.2, hardness = 7.5. Named after Andalusia in Spain.

ANKERITE

A calcium-iron carbonate that varies in colour from white through gray to yellow and brown, ankerite crystals have a pearly lustre and brittle, well-defined rhombohedron habit. Found in dolomite in compact form, and in siderite. Hardness less than 4, SG = 3.1.

APATITE

Nothing to do with dieting; it gets its name from the Greek word 'to deceive,' because it was often confused with other minerals. Occurs in veins with quartz, feldspar and iron ores. Hexagonal crystals, white, green, brown, yellow, even violet. Streaks white. It is used as the standard Mohs hardness of 5. SG = 3.2. A thin chip colours a gas flame orange.

APOPHYLLITE

A complex silica, apophyllite forms beautiful tetragonal crystals, and is found in basalts and tufas in association with *stilbite*. Colour varies from yellow to pale green. Not really suitable for jewelry since it has a hardness of only 4.5 – 5 (SG = 2.3), it makes an interesting collector's piece however with its perfect basal cleavage and clean lines.

ARAGONITE

Aragonite is a form of hard (3.5 – 4), dense (SG = 2.9) calcite. It forms in many different ways. For instance, oysters make pearls out of aragonite, and the bare skeleton of a coral is also aragonite. For the rock hound, these aren't options, but it also forms in orthorhombic crystals, varying in colour from white, to pink and even blue. In its form as alabaster, it is cut into slabs as ornamental stones.

ARGENTITE

The sulphide of silver, and an important silver ore. Found in massive or cubic crystal form. Colour silvery when fresh, black or gray when tarnished. H = 2.5, SG = 7.3. Often found in association with lead and zinc deposits.

ARGILLITE

Argillite (pronounced *are*-jil-ite) is a fine-grained shale that cleaves in thin slabs, and was the "slate" that so many kids of a few generations ago carried to school. It is common throughout the world, and can contain fossils, oil, or even copper. In its dry state, it is hard, but when first quarried, it is soft and easily workable. It is this workability that has made the Chuck Creek Quarry on the Queen Charlotte Islands so famous, as it is from this site that the Haida Indians have taken their material for their dark stone carvings. The mineral is reserved exclusively for the Haida, and cannot be removed from the Queen Charlottes in unworked form. The carvers bury it in the ground to keep it moist and workable until such time as they need a piece.

ASBESTOS

Asbestos is a variety of *serpentine* in the *chrysotile* group of minerals, having no cleavage, a hardness varying from 2.5 – 4, and unmistakable fibres that are elastic enough to be woven, having a silky lustre. The longer the fibre, the more valuable the mineral, since (despite its questionable health record), asbestos still serves as a valuable additive to many manufactured products where heat resistance or strengthening are required. One of the largest deposits in the world is at (surprise!) Asbestos, Quebec.

It is usually found in thin veins filling cracks, and can be identified easily because nothing else is as fibrous. The fibres can be divided into almost invisible strands, and vary in length from about 1 cm to 8 cm.

AXINITE

Axinite occurs commonly in altered calcerous rocks, where it can be either massive (large lumps with very fine grain), or crystalline, in any colour varying from violet and pink to yellow and brown. Transparent crystals (triclinic) are used as gemstones, having a hardness of 7.

BARITE

Sometimes found in limestone cavities, hydrothermal veins, or in sulphides, barite is a variable mineral that can form attractive, thick rhombic crystals in a variety of colours, or can occur in rosettes as the *desert rose*, which is much prized by collectors. Crystals as long as 1 metre have been found in England. It is soft (2.5 – 3.5 hardness) but quite dense at 4.5, it often has a pearly or glassy lustre, but streaks white. It is the main source of barium, which unlucky patients get to drink before being X-rayed. Also used in the paint industry as a pigment.

BASALT

Basalt is the product of extrusive volcanic lava flows that spread out and cool rapidly to form dark, massive, fine grained rocks, made up mostly of *pyroxene* and *plagioclase feldspar*. *Olivine* can also be present. Much of the Western and Central divisions of the province have extensive basalt covers. In colour, it varies from dark gray with a greenish tinge, to black, although in the dry interior, it can be covered in a pale gray or even white crust. On the coast in the rain belt, the iron can sometimes leach out, giving the matrix a rusty look. Underwater, it cools quickly, forming *pillow basalt* which looks, as the name implies, like a pillow or the top of a mushroom. The upper surface of a basalt flow can sometimes be filled with gas bubbles, forming a porous variety called *scoria*. These holes can later be filled with agate or calcite, making interesting samples.

Basalt can also be found as a dark band of rock forming a dyke or sill (there's a famous vertical one on the right hand side of the Squamish Chief, the huge rock wall overhanging the town of Squamish. It's famous because there is a rock climb that follows the entire length of this huge dyke!), or as hexagonal columns, such as at Green Lake, BC, and the Devil's Causeway in Ireland.

BERYL

An ore of beryllium, a minor metal, it is alloyed with copper, and used in nuclear research. Transparent beryl crystals are the stuff of dreams: aquamarine, chrysoberyl, corundum (sapphires and rubies) and emerald, depending on colour and impurities. All are highly prized gemstones. Hardness usually 7.5 – 8, SG = 3.6.

BLOODSTONE

Bloodstone is a variety of *jasper*, and is usually a dark green with red spots of iron oxide, which gives it its name. The more spotty, generally, the more it is prized. Bloodstone is often used in men's signet rings, or as March birthstone rings. It has a hardness of 7, and poor cleavage, so it's ideal for all kinds of jewelry such as cabochons or teardrops.

BORNITE

A copper-iron sulphide, usually massive, it is soft (H = 3) with poor cleavage, and a gray-black streak. A valuable copper ore, its surface tarnishes upon exposure to air to form *peacock rock*, with iridescent blue, green and purple shades. SG = 5.7, rare crystals are cubic or rhombohedric. Often found with malachite.

CALCITE

Calcite can take many forms, such as *microcrystalline* (limestones), *saccharoidal* (marbles), *fibrous* (alabasters), *concretionary* (stalactites and stalagmites), and more. It is often tough soft (hardness = 3) with a density of 2.7, perfect rhombohedral cleavage, and is basically calcium carbonate.

Collectors prize calcite when found in crystal form. In BC, there are several types that occur: *aragonite* is a harder (4) and denser (2.9) variety with no cleavage, usually found around gypsum; *chalk* is a white, soft form, usually full of tiny sea animal shells; *dogtooth spar* is a common form with sharply pointed crystals; *iceland spar* is a transparent crystal which has the property of bending light in two directions, so that when looking through a slab of it, everything below appears double; *mexican onyx* is not true onyx, but a calcite with swirls and lines that resembles onyx (onyx has a hardness of 6.5 while calcite is 3; *travertine* is a massive, fine-grained form found in caves; *tufa* is a porous white form of travertine.

CARNELIAN

Carnelian is a clear red, red-brown or orange *chalcedony*, having a hardness of 7.

CASSITERITE

Tin is a metal used a great deal in foodstuffs, where a thin layer of it coats the iron canister, preventing the iron from rusting and spoiling the contents. Nearly all tin comes from the ore *cassiterite* (tin oxide), which is almost 80% tin. The ore is brown or black, but streaks pale, with a glassy lustre. Sometimes forms crystals, but more often occurs as *wood tin* which looks like a fibrous mass, or as crusts or veins in pegmatites. Hardness 6 – 7, and quite dense with an SG of about 7.

CHABAZITE

Chabazite is a *zeolite*, often found in association with *stilbite*. It forms large rhombohedral crystals that are almost cubic, usually white or pink with a glassy lustre, transparent or translucent. With a hardness of 4 – 5 and an SG of 2.1, a cluster of chabazite crystals is a real find.

CHALCEDONY

Pronounced "cal-*sidney*," it is a major part of the quartz family, since it covers a wide variety of waxy, fine-grained translucent stones of interest to the rock hound. It forms in cracks and seams in volcanics and is widely distributed in BC and, in fact, the whole of Canada.

Some of the types of chalcedony found in BC are: *agate*, a variegated chalcedony; *carnelian*, which is translucent red or orange; *chrysoprase*, a nickel-stained apple-green chalcedony; *onyx*, which is basically an agate with flat layers, rather than circular ones; *plasma*, a faintly translucent chalcedony with white or yellow dots; *sard*, a deep smooth brown colour; *sardonyx*, which is a combination, obviously, of sard and onyx in the form of flat layers of brown and white.

CHALCOPYRITE

Chalcopyrite is the most common copper ore mined for copper. It is often a brassy, almost golden, colour with a hardness of 3.5 – 4, and an SG of 4.2. It streaks greenish-black, and is very brittle. It may form small tetrahedral-like crystals (rare), but is generally found in massive form. Sometimes found with other copper minerals such as *malachite*, or with *pyrite*, making an attractive base for the fool's gold (pyrite) crystals.

CHERT

Duller and more opaque than jasper, chert is generally a tan/brown chalcedony (silica), but may be greenish or other colours. Equally hard, it chips with a conchoidal fracture like flint or obsidian, and was used for early tools.

CHRYSOCOLLA

Found with other copper minerals, green chrysocolla is a copper-quartz compound and has a hardness of between 2 and 4, with a density of 2. Crystals are very rare. It is valued primarily for its attractive green colour, similar to *malachite*, and is found in thin veins, or occasionally, if you're lucky, in grape-like (*botryoidal*) clusters. Although rare in BC, the opening of new copper mines may increase the supply, although it is too soft for jewelry, and is best as a mineral specimen.

CINNIBAR

The major ore of mercury, this mercuric sulfide is associated with low temperature hydrothermal veins and volcanic deposits. Usually found in granular reddish masses with free mercury droplets in the voids, its weight (SG = 8) and softness are recognizable properties.

CONCRETIONS

Concretions are found all over the province, and are known variously as "mud balls" and "clay balls." We believe they were formed by calcite and silica depositing out of groundwater and forming layer upon layer of material, like an onion. They come in every imaginable shape and size (1 cm to 30 cm diameter), and can look like teardrops, donuts, bowls or even human figures. They are found mostly in clay beds alongside rivers, or in soft sandstone beaches, where, being harder than the surrounding rock, they remain when the beach weathers out. Being clay, many dissolve with time in water, but they look great on a shelf, and are the whimsical pride of many rock hounds.

CORUNDUM

Corundum materials are all alumina (Al_2O_3), and are difficult to identify, except that they are very hard (9). In fact, they are the hardest things in BC, as the province is short on diamonds. There are two types. The black or gray varieties are found along the Fraser River, but are difficult to identify, except by hardness. Corundum and magnetite, when crushed, give us *emery*, which is used as an industrial abrasive.

The other type varies in form from clear (sapphire), through red (ruby) to yellow (topaz), blue (sapphire again), to green (emerald). You can quickly figure out that corundum of the coloured variety is rather valuable. Unfortunately, the province has yet to show any of this type, but ... well, that's what you're out there for, right?

The crystals form rhombohedrally, are heavy (SG of 5.2), fragile, with no cleavage and are found in volcanic lavas, pegmatites and hydrothermal vents. Obviously, any group of stones with such a high pedigree commands a lot of respect in gem circles, and history is full of stories of what folks have done to possess these beautiful crystals.

CYANITE

See *kyanite*.

DALLASITE

Dallasite was named for Dallas Road, Victoria, which runs along the southern tip of that city. While not exclusive to that area, it is the provincial capital's "personal" stone in many ways. A volcanic *breccia*, it usually contains well-defined, thin, rectangular chunks of green or brown lava set in a quartz-rich matrix.

DIOPSIDE

A light green or brown pyroxene (silicate), diopside is found in igneous rocks such as dykes and sills. Like most pyroxenes, it cleaves into cubic fragments fairly easily. Diopside is often found in metamorphosed dolomite marbles. Hardness of between 5 and 6, and an SG of 3.4.

EPIDOTE

Epidote is common in metamorphic rock, in cracks and seams, as thin green crusts, or as crystals. It is typically found where igneous rocks have contacted limestones. The crystals tend to be thin and needle-like, green to brown or black, have a hardness of 6 to 7 and a distinct cleavage.

FELDSPAR

Feldspars form the world's most abundant group of minerals, with many forms. They are found in nearly all igneous and metamorphic rocks. In granites, for example, which are made up of quartz, mica/hornblende and feldspar, it is the feldspar that gives the granite its characteristic colour of white, gold or pink, since it makes up as much as 60% of the mix. Next time you walk past a bank or public building faced with granite, take a look at the feldspars — they are very pretty.

FERRIERITE

Ferrierite is a form of *zeolite*, and so is related to the feldspars, with chemically bonded water added.

GALENA

Galena is the main source of mined lead (PbS), and has been smelted since before Roman times. It's usually found in veins in carbonate rocks, together with zinc, copper and silver, so the ore is valuable for several reasons.

Galena is a heavy (SG of 7.5), brittle silver-gray ore which is sometimes formed into beautiful cubic crystals having perfect cleavage. These crystals, incidentally, were used in very early radios ("cat's whiskers" or "crystal sets"). Galena streaks gray, and has a hardness of 2.5.

GARNET

Garnet is best known as a "poor man's ruby" — often a deep red, hard gemstone, but in fact it is common in many igneous and metamorphic rocks. Garnets are a family with very common characteristics, and crystals are abundant; unfortunately, they are usually about the size of a pinhead! With twelve or twentyfour sides, they appear symmetrical and rather like a ball unless examined under a magnifying lens, when their smooth faces are obvious. With a hardness of 7, they are found in schists, gneiss and marbles, and sometimes in lavas and granites. They are often crushed and used as *garnet paper*, like sandpaper.

Members of the garnet clan are: *almandite* is the common red garnet found in metamorphics; *grossularite* may be white or green (the name refers to the Latin name for gooseberry) and usually has traces of chromite; *pyrope* is the precious garnet variety known sometimes as "cape ruby," and is (of course) rare; *spessartite* is a yellow to red garnet with iron that occurs in granites and pegmatites, *uvarovite* is found in serpentines, and is usually green.

One of the main reasons for mentioning garnets is that in other important areas in the world, garnets are found in *kimberlite pipes*, a geological form of igneous intrusive that can contain diamonds. While kimberlites are hard to find, and not every kimberlite has garnets, and not every kimberlite with garnets has diamonds, still ... you can see where we're heading. To date, no diamonds have been found in the province, but BC is huge, and the opportunities vast. Finding a kimberlite could just make your whole day.

GOLD

Heavy, malleable, slow to tarnish, and rare, gold has been the cause of more human disruption, effort, valour and failure than all other minerals rolled into one.

It is soft (with a hardness of 2.5) and is sometimes alloyed with copper to harden it and make it go further. As an alloy, it loses that deep rancid butter colour and becomes paler, like margarine. Pure gold is 24 carats; 14 carat gold is 14/24ths gold, or about 60% pure.

Gold is found in quartz veins, sometimes with pyrite (fool's gold). Being very dense (SG of 19.3) it works its way into rivers and creeks and is deposited in deep cracks and fissures from where you can pan it as *colour*, *flake* or *nugget*.

It would not be untrue to say that the discovery of gold in the 1850s is what created the province of British Columbia, and even today there are new gold mines being opened every decade. But by and large, the move has been away from the placer deposits in streams and rivers, to hard-rock mining, which is the realm of "the big boys." For more information, see chapter 7.

GROSSULARITE

Grossularite is a type of white/green garnet, and is quite common in BC, where it can be confused with jade since it has similar colour and hardness. The surest way to tell them apart is to look at a break — jade has a fibrous break, while grossularite is granular or crystalline — the difference being most obvious under a x10 lens.

To make things more confusing, grossularite is sometimes called "garnet jade" or "Washington jade," or "Oregon jade" (from those respective states). There are various names for different colours, too: *amethystzontes* is purplish-red; *essonite* is cinnamon; *jacinth* is reddish-orange; *landerite* is pink or rose; *succinate* is yellow-amber. Grossularite luminesces strongly under UV light. See also *Garnet.*

GYPSUM

Gypsum is a common mineral, and is the basis for the huge industry of plaster, plasterboard, rock lathe and other building materials. The ancients used to dry gypsum in kilns, then grind it to a powder; today we call it *plaster of Paris*. When water is re-added, it sets hard again.

Gypsum is calcium sulphate, sometimes having impurities that colour its whiteness. With a hardness of just 2, and an SG of 2.3, it has a pearly lustre, and streaks white. *Selenite* is crystalline gypsum, found in limestone cavities; the crystals can grow to 1 metre in length. *Anhydrite* is gypsum without water; crystals are rare. It is harder (3 to 3.5) and denser (SG = 2.9) than gypsum.

HEMATITE OR HAEMATITE

Hematite is an iron sulphide ore, having about 70% iron, which explains its weight (SG of 5). It is an important source of iron ore. Hardness varies from 1 to 6, depending on the ore's chemistry. When tumbled, it takes on a smooth, shiny metallic lustre that is unmistakable, and popular in jewelry. The name comes from its curious property of streaking cherry red, due to iron oxides present. The Greeks thought it was blood, and called it "haema" (blood) stone.

IDOCRASE

Once called *vesuvianite*, idocrase is a glassy, transparent to translucent mineral found in metamorphosed limestone. Found as a crystal, or massive, it is hard (6), with an SG of 3.3. The green variety (*california jade*) is often offered as a jade substitute.

ILMENITE

An iron-titanium oxide, hard and heavy, and generally brown or black in colour. Found often in schists and gneisses, it is weakly magnetic when cool, but becomes more so when heated. A source of titanium (used in hi-tech metals), large concentrations have been found in marine sands.

JADE

Jade gets its name from a corruption of the Spanish name *piedra de ijade*, meaning "stone of the loins." When Cortes ventured into Central America, one of the many wonders he learned was that the Mayas used jade to cure kidney disorders. In fact, like the Chinese and Maori cultures, Montezuma's civilization had based its whole culture on jade, and the story goes that after his first meeting with Cortes, Montezuma was supposed to have said, "Thank God they are only after the gold and silver. They don't know about jade."

Piedra de ijade was translated into French as *"pierre de l'ejade,"* but somehow ended up as "*le jade.*" When the mineral was being classified using the traditional Greek style of the 18th century, it was called *lapis nephriticus*, literally, stone of the kidneys, just like the original Spanish. That's how we get "*nephrite.*"

Jade is the provincial stone of BC, and worthy of the honour. However, to confuse things, jade is the name given to two different minerals that are really mineral *aggregates*. The type of *actinolite* found in BC (and in lesser amounts in China and New Zealand) is *nephrite* and its toughness, hardness (5 – 6) and translucent colouring is what has made it so desirable for centuries, notably in China, where it has been at the centre of a culture going back at least 4,000 years as a tool, weapon, and art form. So, if anyone tells you that BC "jade" isn't *real* jade, tell them it's the same stuff the Chinese sculptors have been carving with such elegance for millennia, thank you.

The other form of jade is *jadeite* (hardness of 6.5 – 7). It belongs to the *pyroxene* group, and first appeared in 1784 from Burma. Some has been found in Washington State. Jadeite is also the stone of choice in Central America, but be careful, much of it isn't jade at all. Despite thousands of years of Chinese excellence carving nephrite, jadeite is today considered more valuable than nephrite, possibly because of its greater rarity. However, from a collector's point of view, the two aggregates are almost indistinguishable (nephrite has an SG of about 3.2, jadeite of 3.4).

Jade comes in just about every colour, from black to white, although that intense green is what most people think about when you mention jade. There is a report that along the Fraser River there is a variety of nephrite known as "chicken innards jade" because of the contorted swirls, if you can imagine pale green guts.

Hunting for jade in BC is usually done along gravel bars at low water. Jade is heavy and fibrous, so tends to collect less mud than other rocks. If washing off the silt on a stone reveals a depth of green colour, the next test is to chip off a sliver (gently, remembering that (a) nephrite is very hard, and (b) a shattered rock is worthless!). The sliver must be translucent — if there's no glow though the edges, drop it and continue. If, on the other hand, it passes that test, the next check is to scratch it with a knife (hardness 5.5). If it passes this, you can start to get excited, and it's time to start thinking about getting back to the workshop and putting it through a slabbing saw. When polished, jade has a vitreous lustre that is unmistakable.

There are lots of fake jades; indeed, you could say the world is awash with the stuff. Green quartz, albite, diopside, aventurine, vesuvianite, chrysoprase, chrysolite and bowenite are all popular substitutes — buyer beware.

JASPER

Jasper is common throughout BC; so common, in fact, that a lot of people lose interest in it. This is a mistake, because jasper comes in such bewildering varieties there's always something new to see. Because it is a quartz (with a hardness of 7, SG of 2.6), it polishes and wears well in jewelry.

Being a *cryptocrystalline quartz* means that it has tiny crystals of other minerals in a quartz mass, so it is opaque, often quite dull in the rough. Jasper is commonly found in smooth red or yellow masses, or with other crystals and chunks in it, when it is known as *breccia* (pronounced "brech-ee-a"). It can be found with fossil shells in it as *turritella jasper*, or with little eye-like dots as *orbicular*. It can have patterns in it as *picture jasper*, dark green with red spots as *bloodstone*; or it can replace wood to form *jasperized wood*.

KYANITE

Kyanite, or cyanite, is an aluminum silicate found generally in schists and gneisses. The crystals are knife-like and long, and can be white to blue-gray, or even black. With an SG of 3.6, it has an unusual hardness — between 4 and 5 along the crystal axis, but 7 across it. It is used to make *mullite* in refractory porcelain and brick.

LAZULITE

Found in azure blue masses, its distinctive colour, vitreous lustre and indistinct cleavage make it easy to recognize, although it can look like lazurite or sodalite. With a hardness of 5.5 – 6 and an SG of 3.1 – 3.4, it's found in pegmatites, quartz veins or quartzite.

LIMESTONE

Most limestones were formed as sedimentary deposits by animal and plant life, to form what is essentially calcite ($CaCO_3$). They vary in colour, form and texture, depending on their origin. The cliffs of the Grand Canyon are good examples of limestone. *Marls* are limestones with lots of clay; with magnesium, it is a *dolomite limestone*. Limestone is the main ingredient of cement, so is an important economic mineral.

MAGNESITE

Generally found in massive or granular form, magnesite is magnesium carbonate which is mined in some places for magnesium, a valuable additive to metal, and used in the aerospace industry. There are large deposits in Washington and California. Hardness of 4, and SG of 3.0.

MAGNETITE

An important iron mineral (Fe_3O_4), magnetite is the only black ore that can be picked with a magnet, and so is easy to identify. (Pyrrhotite, which is also magnetic, is yellow/gold.) It is quite hard at 6, with an SG of 5.2, containing 72% iron by weight.

MALACHITE

Often found with *azurite*, both are copper carbonates. Malachite forms smooth, irregular masses of beautiful green shades and swirls, and is much favoured for jewelry, boxes, pyramids and other ornaments. It has a hardness of 4 and an SG of between 3.7 and 4.

MARBLE

Marble is a recrystallized limestone, usually white, but can be tinted almost any colour due to impurities. Limestones altered by trickling water are often called "marble," but to be a real marble, it must have undergone metamorphosis of heat and pressure. Marbles seldom show banding characteristic of schists and slates. Used extensively in building.

MARCASITE

Has the same chemistry as iron pyrite (fool's gold), and is often called *white pyrite* although it is more brittle and lighter than true pyrites. It forms in radiating or cockscomb forms, as crustations in peat or clay. It streaks gray-brown. Tends to be very crumbly with poor cleavage. Oxidizes quickly in air, forming white flakes of melanterite. SG = 4.3.

MOLYBDENITE

Usually found in veins or pegmatites, this sulphide is very soft with a hardness of 1.5, has a metallic lustre, and streaks gray-blue. Occurs as flecks or tabular crystals. Molybdenum is an essential additive in high strength steels.

MUSCOVITE

Muscovite is a pale, almost colourless mica, named after Muscovy where it was used as a glass substitute. Common in granites and pegmatites, it has a hardness of 2.2, SG = 2.8, and nearly perfect cleavage.

OBSIDIAN

Obsidian, pumice and rhyolite are chemically the same; the difference is in their cooling. Obsidian (sometimes called *volcanic glass*) is a classic extrusive lava that chills quickly, allowing no crystal separation. It is black and glassy, and breaks conchoidally (shell-like cusps). Along with flint, it has been used by primitive cultures for tools, arrowheads and ornaments (hardness = 6). In BC, there was an active trade between the coast and prairie peoples in obsidian long before the white man's arrival, since the stone keeps a sharp, hard edge ideal for knives and spears.

Varieties of obsidian are *snowflake obsidian*, which has small white crystals of devitrified glass in the black matrix; *apache tears* which are translucent round nodules of obsidian found when the main mass has been altered; and *rainbow obsidian* which has a beautiful play of colours.

When collecting obsidian, it's important to wrap it very well, as it can cut its way through newspaper and damage surrounding specimens. It is a very popular rock with collectors, and is used widely in jewelry.

OPAL

Formed by silica (quartz) with between 3% and 9% water, opals are an enigma. They come in many colours and tints, have a glassy or waxy lustre, and have no chemical structure (they are *amorphous*). There are two types of opal; *precious*, and *semi-precious*. Some varieties (notably from Australia) have a striking play of colours called *opalescence*, making them highly desirable.

Opal is porous and easily stained, so avoid immersion in dirty water. For this reason, beware of most brilliantly "coloured" opal, which is in fact dyed stone. Opal loses water easily, and should be immersed from time to time to avoid cracking. Opal should not be exposed to sudden temperature changes either; the stresses can split the material.

Opal looks like quartz, but has a hardness of 5.2 to 6.5 (use a hardened file), so can readily be distinguished. It has an SG of about 2. It is found in both igneous and sedimentary rocks, and often replaces the material in petrified wood, making fine natural sculptures of *opalized wood*. When handling opal, be very, very careful — it's as fragile as it is beautiful.

There are several varieties of opal. *Common opal* is widespread in volcanic rocks, being milky, green, yellow or brick red with a glassy, translucent look; *hyalite* is a clear, colourless variety found encrusting rocks or filling small veins; *geyserite*, as its name implies, is found near hot springs, and is usually white and opaque; *tripolite* is formed from microscopic shells of diatoms, and is chalky, fine-grained and hard enough to scratch glass.

PASTELITE

Pastelite is an agate breccia and, as its name implies, it ranges in colour through the whole spectrum. Being an agate, it is hard (7) and makes attractive cabochons and medallions, highlighting the various patterns in each piece.

PEGMATITE

An important metamorphic (intrusive) material, pegmatites often include well-defined crystals of quartz, feldspar and mica (muscovite, biotite, lepidolite), and may contain tourmaline, beryl, topaz, zircon or other rare materials. These are therefore rocks of economic importance.

PENTLANDITE

This heavy iron-nickel sulfide is the main source of nickel at Sudbury, ON. Its colour is often light bronze-yellow, with a metallic lustre. Streaks bronze-brown. H = 3.5, SG = about 5.

PERIDOT

Also known as *olivine* or *chrysolite*, peridot is the most common member of a group of silicates. Peridot is found wherever igneous rocks are rich in magnesium and poor in quartz, such as basalt or quartz, or in metamorphosed dolomites. Mostly green, with a hardness of 6.5 and an SG of 3.3, clear peridot is cut as a gemstone, where it's prized for its fresh, apple green colour. Crystals are rare.

PETRIFIED WOOD

Many people think petrified wood is wood that has dried out in a desert and been replaced by sand. In fact, the opposite is true. It must first be buried;

then the wood fibres must be slowly replaced by liquid minerals which then harden, taking the shape of the original fibres; finally, the now petrified wood must be exposed again to the forces of erosion.

Petrified wood (from the Greek word *petra*, a stone) is fairly easy to recognize, since it looks like wood, although there are some schists that look very bark-like. The value of a piece depends to a great degree on the detail of replacement, and the mineral that did the replacing. Opal and agate are popular, since they are often spectacular in colour, but calcite, dolomite or pyrite are all known to make up fossilized wood specimens.

PLATINUM

Native platinum and the mineral *cooperite* (PtS) are found in dark extrusive igneous rocks (called *basic* or *ultra-basic* rocks). Platinum, sometimes called "white gold," was definitely the noble metal's ugly sister — in South America the Spanish threw it away when it was found along with gold and silver! It is only in this century that its value as an inert (non-reactive) metal was appreciated. Because it doesn't oxidize or react willingly with anything, it's used widely in laboratories for ultra-pure containers, as a catalyst in industrial chemistry, and to a lesser degree in jewelry.

Like gold, it is malleable, with a density of 14 to 19, depending on purity, but is gray/silver in colour. Hardness of 4 to 4.5, it streaks gray. Crystals (isometric) are very rare (typically 2 mm across), and extremely valuable. Usually found as grains or small nuggets in placer deposits, like gold; because of its specific gravity, it can be panned like gold.

PORPHYRY

A porphyry is any igneous intrusive or extrusive where the rock has a texture of at least 25% coarse grains or crystals (called *phenocrysts*). Thus a granite having at least a quarter of its matrix made up of phenocrysts would be called a *granite porphyry*. Similarly, we could find a *basalt porphyry* or a *syanite porphyry*. The *gabbro porphyries* of Vancouver Island are known as "flowerstone," because the feldspar phenocrysts look like white daisies on a dark (gabbro) background. Very thin phenocrysts are called "mouse track" porphyry.

PREHNITE

See *zeolite*.

PYRITE

Pyrite (FeS_2) or iron pyrite is often called "fool's gold" because of its gleaming gold colour, but gold seldom forms crystals, and pyrite has much greater hardness (6). With an SG of just 5 when compared to gold's 19, it won't fool you for long. Still, a large cluster of pyrite makes an attractive collector's piece, and it is often found with gold in quartz.

PYROLUSITE

Usually found as fibrous or earthy masses, this oxide of manganese can vary in hardness from soft and crumbly to a hardness of 6.5 in crystals, which are rare. It has a greasy feel and streaks black. SG about 5, it is an important ore of manganese. Forms fern-like dendrites in cracks and moss agates.

PYROPHYLLITE

A very soft (1 – 2) silicate, crystals very rare. Usually yellow, white or pale green, it has a greasy feel and pearly lustre. Difficult to distinguish from talc, except chemically.

PYRRHOTITE

An iron-sulphide that frequently has surplus sulphur such that a new break will smell strongly of sulphur. Usually golden/bronze, with an SG of 4.6 and a hardness of 4. Streaks gray-black, and is often magnetic.

REALGAR

Realgar is an arsenic-sulphur compound that forms in low temperature veins as a crust, grain or fleck, or as massive deposits. It is usually red, soft (H = 2), with an SG of 3.5. Crystals are rare. Realgar slowly breaks down if exposed to light, turning into *orpiment*, (As_2S_3).

RHODOCHROSITE

A pink manganese carbonate that is quite soft (H = 4), it is found most often in veins or crusts, except for a spectacular deposit in Argentina. Much used as an ornamental stone, and in jewelry, but has perfect rhombohedral cleavage. SG about 3.5.

RHODONITE

A manganese silicate, rhodonite has a hardness of about 6, and is popular for jewelry and ornaments because of its pale to hot pink colouring, often with strands of black in the matrix. Prismatic cleavage. Crystals are fairly common, and are large and flattened. Derived from the Greek *rhodon*, meaning a rose. When found, it nearly always has a crust of black manganese oxide hiding its inner colours. Sometimes mined for its manganese content.

RHYOLITE

Rhyolite is a light-coloured acidic extrusive igneous rock having the same composition as granite, but, being extrusive, it has cooled more rapidly, so is fine-grained. When more than about 25% of it is quartz phenocrysts, it is called *rhyolite porphyry*. Colour is white to pink to gray, although iron can stain it red in places. Many rhyolites are porous, and can absorb mineral-rich solutions that infiltrate the rock, causing attractive banding. When the infiltration hardens the base material, it is called *wonderstone*, and can be made into unusual jewelry similar to picture jasper.

SARD

See *carnelian*.

SARDONYX

Sardonyx is a form of banded chalcedony (*onyx*) with even, parallel bands of black and white, or brown and white.

SCHEELITE

Scheelite is an important tungsten ore that is found in quartz veins or at the contact of igneous rocks and limestone. Colour is variable, but usually light. Hardness of 5, SG = 6, with a glassy, sometimes transparent lustre. Streaks white. Good cleavage, tetragonal crystals. Fluoresces in short wave UV-light.

SELENITE

See *gypsum*.

SHATTUCKITE

This rare hydrous silicate of copper closely resembles blue azurite, and quality samples are used in a similar manner. Crystals are rare; usually found in irregular masses of spherules with tiny crystals radiating from a central point. This separates it from azurite. SG = 3.8, hardness variable between 3.5 and 4. Usually associated with azurite, malachite and chrysocolla.

SIDERITE

Siderite is an iron-carbonate, but the iron content is low at 48%. However, because it is sulfur-free and often has manganese associated with it, it is mined. Rhombohedral crystals are common, or sometimes appear as botroidal clusters. Found in hydrothermal vents, quartz vugs, associated with galena and sphalerite. Colour yellow, brown or gray. Hardness about 4, SG = 3.8. Streaks white. Lustre pearly.

SILLIMANITE

This aluminum silicate is usually silky/fibrous in shape with long, slender crystals, having a pearly luster. Hard (6) and heavy with an uneven fracture. Found in metamorphic rocks, and used in the refractory industry.

SILVER

Second only to gold in history as a noble metal, silver can be found as native metal in large, twisting, branching masses, or as silver sulphide — *argentite* (from the Latin "argentum" meaning silver or money). Shiny when clean, silver tarnishes fairly easily. Argentite may be massive, or can form cubic crystals; hardness 2.5, SG of 7.3, with a metallic lustre. Often found together with lead and zinc and other rare minerals.

SKARN

A dark metamorphic rock containing calcite, pyroxene, garnet, sulphides. Often the host rock for minerals containing copper, iron, manganese and molybdenum.

SOAPSTONE

Soapstone is basically talc and colouring, and so is soft enough to be scratched with a fingernail. It is ideal for carving, and is the preferred medium of the arctic Inuit, and is much in demand. It can be found in many colours, although a mottled brown is the least attractive and the most common. Green soapstone is very valuable. Easy to identify, soapstone feels soapy as well as being very soft.

SODALITE

Sodalite is a bright marine blue silicate, with an SG of 2.3 and a hardness of 6, making it popular as a semi-precious stone. The sodalite capitol of Canada is Bancroft, Ontario, but there is quite a bit in BC too, although most of it to date has been found in the national parks, where collecting is prohibited. Seldom found as a crystal, the blue can sometimes be tinged with green.

SPINEL

As a gemstone, it was confused for centuries with ruby, being very hard (H = 8). Brown, green and blue have been found, depending on elements present. Non-gem spinel is found in association with marly dolomites. SG varies from 3.5 to 4.1.

STIBNITE

Stibnite is the sulphide of antimony, usually found with pyrite, galena or arsenic minerals, usually in low temperature veins. Long thin steel-gray crystals are quite common, sometimes bent. Soft (2) with perfect lengthwise cleavage. SG = 4.6. Antimony is used in low friction metal alloys, pewter, and in batteries.

STILBITE

Stilbite is a *zeolite* that forms in wheatsheaf-like crystals of white to yellow to reddish-brown. Hardness between 3.5 and 4, SG of 2.1. Often found in lavas, filling cavities and veins.

TALC

This hydrous magnesium silicate never occurs in distinct crystals, but instead in scaly aggregates of white, gray or brown. Has a greasy feel. Very soft (H = 1) with perfect cleavage. Can be worked with hand tools as *soapstone*, and is commonly found with schists. Used in paper, rubber, paint and, of course, cosmetics as talcum powder.

THOMSONITE

This is one of the *zeolite* family, found in metamorphics, lining vugs and veins. Usually blue and botryoidal (like a bunch of grapes). Because of its softness (only 2.5 to 3) it is not jewelry material, but favoured as a collector's piece.

TOURMALINE

Tourmaline is a complex aluminum silicate that is occasionally common with mica and feldspar in granite rocks and pegmatites. It forms long, needle-like crystals with striated sides and very distinctive triangular ends. In BC it is often black (*schorlite*) due to iron impurities, and has little value. Elsewhere, two-coloured crystals are found, with one end of the crystal being green, the other rose, red or orange (called "colour zoning"), when it's called *watermelon tourmaline*. Streaks white; hardness of 7, SG of 3, with a vitreous lustre.

TRAVERTINE

Travertine is a massive, non-crystalline form of *calcite*, often found in caves. Opaque, often coloured with impurities.

TUFF

Tuff is volcanic ash, often having a variety of small and large angular fragments. May be layered like sedimentary rocks, although it is of igneous origin. Large deposits have been found considerable distances from volcanoes. Colour usually pale gray or brown. Sometimes used as a cheap construction stone.

TURQUOISE

Turquoise is a secondary mineral formed in dry regions from the alteration of aluminum-rich rocks. It is a waxy, vitreous mass that varies in colour from blue-green to light green, and has been highly prized from ancient times for jewelry. Having a hardness of between 5 and 6, it is distinguishable from "me-too" copies such as chrysocolla (H = 2 to 4) and variscite (H = 3.5 to 4.5).

Much valued by native North Americans, turquoise and silver jewelry is popular and commands high prices. But beware of imitations — turquoise is often dyed to appear more blue, and there are many imitators!

THUNDEREGGS

While Oregon state is famous for its thundereggs, these unusual forms of agate, jasper or opal are found as far north as the southern edge of BC. Ranging in size from 1 cm to as much as 1 metre, thundereggs appear from the outside as a nodule of warty rhyolite that looks anything but interesting. But cut it in half, and a stunning void filled with startling patterns of chalcedony appear, each unique and much in demand by collectors. Many thundereggs have an axis of symmetry, meaning that if you cut along that plane, the left and right hand pieces will closely mirror each other. To guess this axis, look for a knob on the outside which has a dip directly opposite on the nodule — that's your best bet for a mirror set of eggs.

VESUVIANITE

See *idocrase*.

WOLLASTONITE

Wollastonite usually forms as fibrous or radiating gray masses with a pearly luster. SG of 4.5 with perfect cleavage.

URANINITE

Uraninite is a steel black, opaque, uranium compound with a hardness of 5.5 and a density of 9 to 9.5. In BC, usually found in sand. Cubic form is very rare, radioactive.

ZEOLITE

Widely distributed, zeolites are chemically related to *feldspars*, with the addition of water. When heated in a flame, zeolites boil and bubble — hence their name, which means "boiling stone." There are about twentyfive in the group, and they are found in lavas, filling cracks, veins and cavities. Three of the more common found in BC are: *chabazite*, having rhombohedral (almost cubic) crystals; *ferrierite*, which display sunburst formations of yellow or pale brown crystals; *stilbite*, having pearly, wheatsheaf-like masses of twinned crystals in white, yellow or brown. All three are pale, soft (3.5 - 5) minerals of low density (2.1).

ZIRCON

Zircon ($ZrSiO_4$) is common in igneous rocks, but fairly rare as a transparent gemstone. When found as a brown tetragonal crystal with a hardness of 7.5 and an SG of 4.7, it can be heated to turn it into a more valuable blue gem. Zircon is also man-made in large quantities in clear form, where its hardness and brilliance can be mistaken for a diamond. It has an SG of 4.7.

CHAPTER 7: GOLD!

Gold, man's first folly

– Pliny, 79 AD

Gold has some important qualities beyond the warm yellow colour and metallic lustre. For a start, it is very rare. You should be aware that gold accumulates too — very little gets lost once it's found (despite all the stories about wrecks on the Spanish Main, and Blackbeard's treasure). Gold tends to be recycled and re-used over and over again, especially in times of strife, so that the ring on your finger could well contain gold worked by Incan or Egyptian goldsmiths.

Gold is the most malleable of metals and can be hammered into a foil a thousand times thinner than a sheet of paper — so thin, in fact, that light can pass though it. For example, it serves as a coating on astronaut visors to shield harmful rays. Gold is also extremely ductile, meaning that a piece the size of a golf ball can be drawn into a wire so fine that it could stretch from Vancouver to the Alberta border (600 km).

These properties, combined with gold's great corrosion resistance, have made the metal vital in many high technology processes such as electronics, aerospace, and medicine. It has always been in demand in the arts, in weaving, ceramics, sculpture, pottery and of course in jewelry. Finally, it has a value as exchange (money). The person who put England (and ultimately, the world) on the Gold Standard was the same man who invented calculus — Sir Isaac Newton. After leaving Cambridge University, where he was Master of Trinity College, he was made Director of the Royal Mint. There, he set out to stabilize the pound sterling, at a time when international trade was just beginning.

Gold has gripped the mind of man the way no other mineral ever has. In the first century B.C., Diodorus Siculus wrote, "*nature herself makes it clear that the production of gold is laborious, the guarding of it difficult, the zest for it great, and its use balanced between pleasure and pain.*" It is hardly surprising that Shakespeare wrote of "*saint-seducing gold*" in Romeo & Juliet!

Canada's earliest gold mine was founded in 1847 in Quebec. The Klondike gold rush began in 1896, and the largest nugget ever found there weighed a staggering 2.64 kg (that's 5.8 lb.)! But that bonanza was nothing compared to the largest ever found, at Carson Hill, California, where a 72.8 kg monster (160 lb.) was discovered.

7.1 RECOGNIZING GOLD

There are two common types of gold found in the province of British Columbia: quartz gold and placer gold. It has also been found sometimes with copper deposits, and in skarns. To the two major sources, it's possible we should add "fools gold" (iron pyrite). Quartz gold is bonded to the quartz, requires big bucks and even bigger machinery to hack, crunch, separate and smelt, so is of little interest here. Placer gold, on the other hand, involves panning in rivers and old stream beds to find gravity-fed gold, that looks like gold, smell likes gold, tastes like gold, and by golly, it is gold. The only way you can be fooled is, of course, by fool's gold. But the differences are relatively obvious:

	Gold	Pyrite
Hardness	2.5	6 – 7
SG	19.3	5
Malleable	bends easily	brittle, shatters
Colour	rich yellow	paler, brassy
Surface	matte, lumpy	smooth crystal surfaces

7.2 PANNING FOR GOLD

There are some excellent books dealing with this subject so we will stick to the basics. A gold pan is a cheap investment and can be bought at many outfitters or hardware stores. Choose a plastic one — they're easier for a beginner, as they require no maintenance (don't rust), are light, stay warm in cold rivers (don't freeze your fingers as much), and allow you to see the gold better. Dark green provides the best contrast; black is also OK.

If you're new to the game, it's a good idea to get some practice before you head out. Get a few ball-bearing sized pieces of lead (cut up a piece of lead sheet, or get hold of some lead shot). Fill your pan to the brim with gravel and sand (easily found at the beach or in some stony gardens), and add the lead shot.

Have running water available (a garden tap is fine). Fill the pan with water. Break up the lumps of soil so the result is a muddy mess (nobody promised you'd get rich AND stay clean!), and then, holding the pan on either side, shake the pan horizontally from side to side or in a circular motion, so that the gravels and soils are thoroughly swirled around. This is the most important part of the panning process. The swirling motion liquefies the mixture, or slurry, in the pan, so the heavy objects can sink through the slop to the bottom. With any luck, the gold (or lead in this case) will do just that.

After about 20 seconds of vigorous circular motion, you'll notice that the pebbles have floated to the top of the slop (yes, pebbles can float, but only on denser mud and gravels). That's a sure sign you're doing it right. If the stones rise, the gold has sunk.

Scrape all the loose stones off the top, refill with water, and repeat until most of the bigger, lighter stones have risen to the top. Refill with water again, and tip the pan away from you, still holding on with a hand on either side. If your pan has riffles (little grooves on one side of the pan's sloping surface), these should be away from you.

Now swirl the pan in a horizontal, rotating motion again, so that each time you swirl, a small amount of the gravel and grit slips over the far edge of the pan, together with some water. Refill from the tap, if necessary; it is important to keep the pan moist so that any gold in the pan settles in the lowest point of the pan, which is the far corner, below the lip that is spilling. Gradually continue to reduce the contents of the pan, swishing around until only a skim of sand and gravel remains in the bottom curve of the pan.

Tip level, and examine it closely; the lead shot should all be concentrated in the pan. If it's not, you're either not shaking hard enough at the start, or are panning too vigorously during the concentrating stage. Try again, until you get every piece.

7.3 WHERE TO LOOK

Gold is heavy. That's the golden rule. Because of this, gold will always sink down through the gravels in a river, wedging itself in the bottom, or in cracks and fissures in the bedrock. It will also collect below sills, waterfalls, downstream of boulders, anywhere where eddies cause a sudden drop in water speed, thereby allowing the gold to fall.

Remember that creeks and rivers change their courses, so be on the lookout for old stream beds; they are often where the real paydirt is! Old river beds (and young ones) can also get covered with a layer of hard clay (often blue gray in colour) that looks like bedrock, and isn't. The real bedrock may be further down.

There is an interesting story about Billy Barker (of Barkerville fame). Bill and his buddies couldn't get a stake on Williams Creek where all the action was, so they headed to a spot below the gorge to try their luck, although the area had already been scoured with disappointing results. At the gravel layer they found the usual streak of dull, dark gold. Unlike the sourdoughs on the upper reaches of the river, they kept going down — 40 feet, then 50 feet. Barker was the laughing stock of the Cariboo, but he refused to give up. At 52 feet they hit bedrock, and with it, the most incredible pay streak. The rest is history — a single pan yielded $5, and a foot of ground turned up an incredible $1,000. By the end of 1862, almost $2 million worth of gold (at $20 per ounce) had come from the Cariboo fields, much of it from Williams Creek. A year later, there were almost 4,000 people living in "Barkerville," where the year before there had been nothing!

When panning a river, gold tends to travel in straight lines, and to cut corners where the river bends. Avoid panning where the river is fast flowing; it's likely the gold will have been carried on by. Pan the soil taken from overturned tree roots, or river banks. Moss is another great trapper of gold. Some of the best finds have been found in these unlikely places.

Stay out of potholes — it's true that gold will drop into them and concentrate, but the pounding water plus the rolling action of the gravels will turn any gold to flour in a single spring flood. Stick to fast flowing rivers that change their speeds; once they are on the flat, they are too slow to carry gold very far.

Finally, be aware that despite all the evidence to the contrary, you are not the first! A century ago, men made of stern stuff moved mountains with shovels and sweat, and believe me, they didn't leave much unworked! In particular, the patient Chinese were known to work an area with a thoroughness that was legendary!

On the other hand, every winter exposes new surfaces, brings new material down into the creeks, and opens up new possibilities. And isn't that what it's all about?

7.4 EQUIPMENT

Gold panning requires pretty much the same equipment as rockhounding, but there are a few differences. In addition to a shovel, pick and prybar (and a pan, of course), you should take along a pair of tweezers for lifting gold out of narrow cracks), a stiff brush or paintbrush (for sweeping gravels out of same), a magnet (to confirm magnetite — a good sign, since it's associated with gold bearing gravels), and a plastic bottle measuring about 2 cm diameter by 5 cm tall (a clear film container is OK, although a screw cap is better) filled with water, to store the nuggets, flakes and flour. Waterproof gloves and boots will allow you to work freezing streams long after the rest of the party have chilled out.

It's often a good idea to pan your gravels down to the concentrate stage (a skim at the bottom of the pan), and then to save that in a large plastic jar with a screw cap, to be panned carefully later, when there's more time, fewer distractions, warmer water, no bugs, and better light.

CHAPTER 8: OTHER TREASURES

8.1 FOSSILS

Fossil collecting requires quite a different mindset from collecting minerals and gemstones. The latter speak for themselves with their colours, shapes, and properties. Not so fossils. On the face of it, a fossil is a pretty ordinary thing to look at — the shape is simple, often indistinct, the colouring dull, and the properties ho-hum. Yet fossils hold some people's attention like nothing else. To hold a fossil, is to glimpse, however dimly, the immensity of time, and the history of life. No doubt about it; fossils appeal to the intellect.

Fossils are most abundant in sedimentary rocks such as shale, sandstone and limestone. They are hard to recognize in metamorphics, and are absent (for obvious reasons!) in igneous rocks. If you can find an unaltered bed of sedimentaries, there is the opportunity to find fossils lying where they lived and died millions of years ago. To be preserved, they had to be covered quickly and fossilized slowly.

8.1.1 Collecting fossils

What can you take? Fossils are now protected under the Heritage Conservation Act. To acquire title to fossils on Crown Lands, you have to apply to your nearest regional office of BC Lands. In theory, this applies to all fossils. In practice, common fossils are seldom assessed. But if you turn up fossils while digging, you are required to notify the Archaeology Branch of BC Lands in Victoria. They may assign someone to check out your find.

8.2 METEORITES

Meteorites are rocks that have fallen from outer space; as a result, they can be found anywhere, and vary in size from a pin head to several tonnes. Large ones have left huge craters as witness of their awesome impact. For instance, the crater on the Ungava Peninsula in northern Quebec measures over 3 kilometres across and is 250 metres deep.

Most meteorites are probably from *asteroids*, and as such, give us valuable information on the geology beyond our little planet. There are three types: *stony*, *iron*, and *stony-iron*. The stony's (*aerolites*) consist mostly of silicates, and are hard to recognize since they look pretty much like any other rock to the casual observer. They are usually more or less round in shape and have a fine-grained texture inside.

The irons (or *siderites*) contain metallic iron compounds and are slightly to strongly magnetic. They may also have cobalt, copper, and surprisingly often, diamonds! They are easier to recognize as they have a fused crust which is dimpled as though pushed in by thumbs, caused by the heat of atmospheric entry. They, too, exhibit a dull brown to black surface, but may have a light interior. Meteorites can be mistaken for weathered rocks or metal slag, but weathered rocks lack the fused look, and slag has a low SG, whereas the SG of meteorites is generally 7.5 with a hardness between 4 and 5.

The stony-irons (*siderolites*) are an equal mixture of silicate material and nickel-iron.

Meteorites aren't something you actually look for; rather, you just happen to recognize one while you're hunting for something else. Very few meteorites have ever been found in BC. Not, we suspect, because there aren't any, but because the terrain is so difficult. By contrast, the prairie provinces have produced quite a few, which tells you something about the Alberta-Manitoba topography. The National Meteorite Collection at the Geological Survey of Canada in Ottawa pays a reward for meteorites. Check Chapter 10 for the address.

The largest siderite on record is the Hoba Iron, which weighs about 60 tonnes, and lies where it was found in 1920, in Namibia, Africa. The Ahnighito Iron weighs 36 tonnes, and was transported by Admiral Peary in 1897 from south of Thule, Greenland to the Hayden Planetarium in New York, NY. By comparison, the largest aerolite to date is the Norton Achondrite, weighing just under a tonne, found in 1948 in Norton, Kansas. The Great Mosque in Mecca, the holiest of holies for the Muslim faith, is built around the Black Stone of Islam, which is believed to be a meteorite of considerable size.

Tektites are small, bottle-green to black siliceous glass bodies that resemble obsidian (volcanic glass), but differ chemically from terrestrial glasses. They have been found shaped like buttons, teardrops, dumbbells, pears and spindles, with their surfaces characteristically pitted due to rapid cooling. Tektites are actually chunks of earth that were thrown into the atmosphere during a meteorite collision. Tremendous heat of impact caused them to fuse and cool rapidly. Most commercially available tektites are found in China, the Czech Republic and the south-central USA.

CHAPTER 9: PROSPECTING SITES

9.1 HOW TO READ A SITE DESCRIPTION

Nearly all the sites described in this field guide are described more fully in MINFILE, which is a computer data set that records many (if not all) mineral sites in the province. If you are serious about finding out more about mineral localities, or armchair prospecting without the hassle of bugs or freezing rain, and computers don't scare you, then go to the back of this guide and send off for the MINFILE program and data disk(s).

Of course, you can't carry a computer around with you in the field, nor can you light fires with it, or roll cigarettes in it, or clean up behind the dog with it, which is the advantage of a field guide (although, not *this* field guide!). But to make it easy for you to switch between the computer program and this guide, many of the conventions are the same.

9.2 THE SITE NUMBER

This is at the top of the site description. The Geological Survey Branch of the BC government has a unique computer file number for most mineral sites in the province. Wherever possible, this is shown so you can reference sites more precisely when visiting your local Gold Commissioner's Office, or District Geologist's Office (see Chapter 10 for addresses).

The first four digits serve as the map number in the 1:250,000 series, so that, for example, the MINFILE number 092B 094 means you will find the site on the federal or provincial map number 092B (which happens to be the Victoria area).

Sometimes, there are two blanks after the first 4 digits (e.g.: 092B 094); other times, these spaces have two letters (e.g.: 092HNW076). This tells you that the mineral site is on map 092H, in the north-west (NW) quadrant. You can often buy 1:100,000 maps of these areas, that give you greater detail. In this case, ask for map 092HNW, which happens to cover Yale.

Crown Publications in Victoria (see Chapter 10 again) sells an inexpensive MINFILE blueprint map (no pretty multi-colours) with all the mineral sites shown in place. Cost in 1997 was just $6, and is extremely useful for people who are serious about exploring an area fully.

The final three digits identify each unique site. Within an area, the sites are listed by increasing number; there is no logic in how they were originally numbered, so that two adjacent sites may have quite different numbers. If these three digits start with an "X" (e.g.: X02), then the site is not yet in the MINFILE data base.

9.3 THE STAR RATING

Find this at the top right corner of the site description. To make it easier, each site is rated: *, ** or ***. This rating is highly subjective! Sites get worked over, and you may find different conditions. Anything marked *** means it is well worth a visit; those marked * are more likely of interest to the more experienced and discriminating rock hound who has already "done" the more popular spots. Also, you will find that *** close to a city is relative, compared to the same designation in the back-country. Generally speaking, sites around Vancouver and Victoria are not that interesting geologically; on the other hand, they are close to home, often just half an hour's drive. For this reason, local sites may warrant a *** partly because of closeness, or convenience, or even because they offer a bonus (a great view, perhaps, or a pleasant walk).

If you are just starting out, try the *** and ** sites first. As your experience increases and your curiosity for rarer or more unusual pieces increases, try the * sites, where you may have to go further, look harder, or dig deeper to find anything!

9.4 SITE NAME(S)

This is found below the site number. Often, sites have been worked for many years under various names. The name line gives the most common or historically relevant name to identify a specific site.

9.5 GEOLOGY, MINERALS, FOSSILS, GEMSTONES

Find this at the top right hand corner of the description. Each site has been categorized into one (or more) of these four. Thus, a 'geology, minerals' site is primarily of interest as a geological example of some rock form, but may, as a bonus, have some collectable minerals. A 'gemstones, geology' classification indicates that the site is primarily of interest as a source of gems, but may also have some interesting geological features.

A "**geology**" site either has very little material to hound, but is interesting to look at, or it may have lots of material, but is in an Indian Reserve, or a National/Provincial Park, or some other closed area, in which case you cannot collect. Nevertheless, if you are in the vicinity, and are interested in form and structure, a "geology" site is well worth a visit.

A **"minerals"** site tells you that there is something to collect (if you can find it). Specimens are usually of interest to the rock collector, and would be collectable. Typical 'minerals' would be anything from rare uranium sand, cinnabar, and kunzite, (getting ** or *** ratings) through to common pyrite, cloudy quartz, skarn or pyrrhotite (likely a *).

"Fossils" are self explanatory.

"Gemstones" include anything that can subsequently be worked into jewelry or other art forms. Typical gemstones are free gold, jade, jasper, rhodonite or agate.

9.6 STATUS

The level of development of a mineral site is described as being in one of five stages:

Showing: there is little mineralization, or very little exploration has taken place to prove larger bodies.

Prospect: some limited work has been done (some survey work, or trenching); may have a track to it.

Developed prospect: site has been developed so there is a reasonably detailed estimate of its value (likely a few drill holes, bulk test samples, and a rough road).

Producer: either open pit or underground, from which ore is currently being mined for commercial gain. Usually closed to all but the most creative rock hound.

Past producer: not currently being mined, but there is usually good access, and interesting trenches, dumps and other features. Beware of old shafts and adits!

9.7 NTS MAP

Gives the National Topographical Series map number. Usually cross references the site number mentioned above. Be aware that a decent map is worth its weight in ... well, gold. A $10 map can save your $200 boots and your $500 blisters and your $1,000 headache. It's a good investment. The sketch maps provided with the sites are there for information only. They are not to scale. They are not perfect. And they do not show everything. A good map does. Work with the two for the best results, especially when you are out in the boonies, or in unfamiliar territory.

9.8 LATITUDE, LONGITUDE

Allows you to pin-point a site using the latitude and longitude coordinates, expressed in degrees, minutes, seconds format. With the arrival of cheap

Global Positioning System (GPS) receivers, you can suddenly afford to know exactly where you are in latitude, longitude and altitude, to within 100 metres, sometimes even better than this. If you are serious about rockhounding, and going beyond the ordinary weekend trip, a GPS will save you a lot of time, and bushwacking, and devil's club, and mosquitoes, and ... you get the picture. The only qualification is that GPS receivers do not work too well in forests or narrow canyons, where the view to the sky is limited or blocked. In those circumstances, a good map, compass, and common sense are still the best way to find your position.

9.9 UTM NORTHING, EASTING

The Universal Transverse Mercator is the alternate coordinate system used world-wide, and appears as pale print at the edge of most colour maps of the province. It is easier to locate a site using UTM, if you are used to the system, because the spacing is the same everywhere, and is metric. Because UTM assumes a 'cylindrical' world which can be projected onto flat maps, there is some overlap, but it is pretty minimal.

9.10 ELEVATION

The estimated height in metres above sea level (if you think in feet, multiply this number by 3.3). The elevation reading is useful in mountainous terrains, since it gives you a contour to follow to pinpoint the site, if the latitude and longitude readings appear unreliable.

9.11 COMMENTS

Notes on directions, points of interest and other miscellaneous information.

9.12 COMMODITIES

Indicates the site's end product. For example, the location may show galena, but the commodity would be lead. Similarly, chalcopyrite ore means the commodity is copper. Commodities are listed in decreasing order of importance, based on economic significance to the mining industry. The recreational geologist may value secondary commodities more. For example, where a site lists copper, lead, zinc, you may find that all three are in massive chalcopyrite and galena, which are of little interest to a rock hound. On the other hand, the associated minerals (see below) offer great finds.

9.13 MINERALS SIGNIFICANT

A list of what to look for in ore bodies, primary minerals, etc.

9.14 MINERALS ASSOCIATED

Secondary minerals that may or may not require detailed care to find.

VICTORIA

The City of Victoria and surrounding suburbs offer only limited opportunities for prospecting. The area is underlain by gneisses and unconsolidated sediments, and intruded by granodiorite and feldspar porphyries in the south.

Limestone is found around Esquimalt Harbour, and is the only mineralization of note. By contrast, the beaches offer an eclectic variety of minerals, brought down from distant sources by glacial action millenia ago. Each winter storm brings in a new selection.

Much of the Saanich Peninsula is overlain with sands and gravels (Butler's Quarry on Keating Cross Rd, for example) deposited there when the Cowichan Valley glacier pushed into Georgia Strait and then retreated after the last Ice Age 18,000 years ago.

The very north tip of the peninsula (Lands End Rd from Coal Pt, Deep Cove and Moses Pt on the west to Swartz Bay Ferry Terminal, Canoe Cove and Curteis Pt on the east) is part of the Comox Formation in the Nanaimo Group, comprising large sedimentary beds, tilted steeply (which makes collecting difficult) and dotted with marine fossils.

At locales such as Hornby Island (Sites 69 – 70) and around Comox, the bedding planes are level, and major discoveries have been made. The Courtney Museum has world-class fossils, including a 16 metre elasmosaur (sea lizard).

The sedimentary beds of the Nanaimo Group had an important influence on the development of Vancouver Island in recent history. It was in this deposit that coal was discovered, first at Port Hardy in 1836, and later at Nanaimo in 1848. Remember that in those days, coal was the fuel that powered an empire.

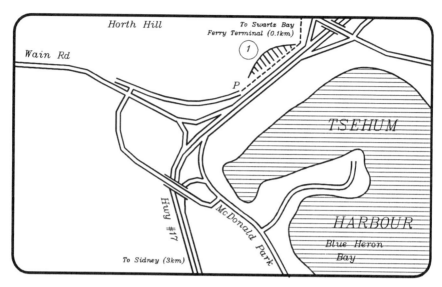

1. MINFILE NUMBER: 092B X07
NAME(S): TSEHUM HARBOUR

* Geology

Status:	Showing		
NTS Map:	092B 11 W	UTM Zone:	10
Latitude:	48 41 00	Northing:	5391800
Longitude:	123 25 15	Easting:	469000
Elevation:	15 m		
Comments:	The northern tip of the Saanich Penninsula, steeply sloping.		
Commodities:	Nanaimo Group sandstones		

MINERALS

Significant:	Sandstone, siltstone, shale, basalt
Associated:	Pillow lava, calcite

The northern tip of the Saanich Peninsula (where you arrive at Swartz Bay on the BC Ferry) shows typical sedimentary layers that stretch the length of the E side (Georgia Strait) of the island. Here in the S, these Comox Formation strata dip steeply at 70 degrees to the N because of the collision of the Pacific Rim Terrane with the Wrangellia Terrane. (That's the offshore plate that is driving in under the Sooke-Port Renfrew coastline.) The Comox Formation, part of the Nanaimo Group, formed when erosion deposited large amounts of material from the mountains along the centre of Vancouver Island during the late Cretaceous era (66 million years ago).

Fossils are common in certain strata, but the tilted nature of the beds at the N end of the Saanich Peninsula and on the Gulf Islands (Piers, Coal, Gooch) makes it hard work to hunt for them. Other areas are along Lands End Rd W of the ferry terminal (check out the road cuttings) and along the beaches. Access to the beach is best where West Saanich Rd runs N down to the sea. Descend the steep bank, and head left or right.

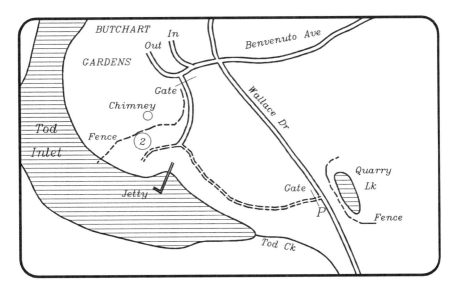

2. MINFILE NUMBER: 092B 006   ** Geology
NAME(S): TOD INLET, BRENTWOOD, BUTCHART GARDENS

Status:	Past producer, open pit		
NTS Map:	092B11W	UTM Zone:	10
Latitude:	48 33 58	Northing:	5379120
Longitude:	123 28 19	Easting:	465190
Elevation:	10 m		

Comments: Centre of flooded quarry at Butchart Gardens. From Victoria, drive Pat Bay Highway #17 to Swartz Bay ferry terminal, and follow the Butchart Gardens signs. There is an entrance fee to the Gardens (well worth it), or visit Tod Inlet by turning S on Wallace Drive off Benvenuto Road. Park 400 m on, next to gate marked "Gowland-Tod Provincial Park." Walk 1 km to Inlet. Lots of limestone in the old walls and ramparts above the jetty.

Commodities: Limestone

MINERALS
Significant: Calcite
Associated: Dolomite

Limestone was produced from three quarries in this area between 1905 and 1921. The deposit comprises a series of limestone bodies that are up to 76 m in width and 150 m in length, hosted in greenstone and intruded by mafic dykes. They generally consist of fine-grained, dark bluish gray to white high calcium limestone. Some 837 thousand tonnes were produced between 1905 and 1921. The numbers 1 and 2 quarries were largely exhausted. The Butchart Gardens presently encompass them. The number 3 quarry is 1 km E, opposite the park gate on Wallace Drive, and is flooded and fenced.

On the W shore of Tod Inlet (Willis Pt) a 6m to 30 m thick calcium limestone bed striking 150 degrees for at least 45 m remains undeveloped.

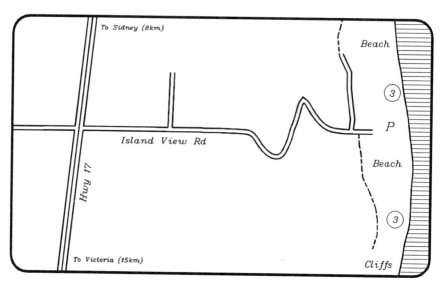

3.   MINFILE NUMBER: 092B 124   **** Gemstones**
       NAME(S): ISLAND VIEW BEACH

Status:	Showing		
NTS Map:	092B 11W	UTM Zone:	10
Latitude:	48 34 28	Northing:	5380000
Longitude:	123 21 58	Easting:	473000
Elevation:	1 m		
Comments:	Located at Island View beach, on the western shore of Saanich Peninsula.		
Commodities:	Gemstones, agate		

MINERALS

Significant: Jasper, agate

Dallasite, epidote, jasper and agate are reported to occur with the beach sediments at Island View Beach. Flowerstone (feldspar) porphyry may be found below Cowichan Head, the high clay cliffs to the S of the parking lot. The cliffs themselves are formed of unconsolidated glacial till, and are the result of ice sheets pushing across the island from the Cowichan Valley meeting the major glacier moving S down Georgia Strait, just 20,000 years ago. Other glacial tills are visibile across Cordova Strait on James Island, notably the large sandy cliffs at the S end.

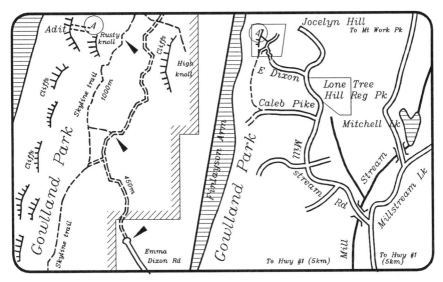

4. MINFILE NUMBER: 092B 098
NAME(S): KINKAM, MERYL (L.90)

** Minerals

Status:	Showing		
NTS Map:	092B12E	UTM Zone:	10
Latitude:	48 32 00	Northing:	5375495
Longitude:	123 31 36	Easting:	461118
Elevation:	350 m		
Comments:	Near the Skyline Trail in the Gowlland Park. Quickest approach is via Emma Dixon Rd, but involves crossing private land. Better route is via Caleb Pike Rd.		
Commodities:	Copper, molybdenum, titanium		

MINERALS

Significant: Pyrite, pyrrhotite, ilmenite, chalcopyrite, molybdenite

Two large parallel shear zones occur, extending about 900 m E from the shore of Finlayson Arm and dipping steeply to the S. The zones contain disseminated pyrite, ilmenite and pyrrhotite, with traces of copper and molybdenum mineralization.

The Skyline Trail crosses a rusty knoll some 600 m S of Jocelyn Hill. To the W, short steep cliffs can be avoided. Drop about 100 m down towards Finalyson Arm. Adit is below tall trees, in cliff.

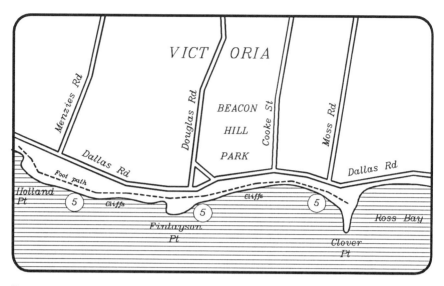

5. MINFILE NUMBER: 092B X01   *** Gemstones
NAME(S): DALLAS ROAD BEACH

Status:	Showing		
NTS Map:	092B 11W	UTM Zone:	10
Latitude:	48 24 25	Northing:	5461000
Longitude:	123 21 40	Easting:	47000
Elevation:	1 m		
Comments:	Located on the beach below Dallas Road, at Mile 0 in Victoria.		
Commodities:	Dallasite (volcanic breccia), porphyry (flowerstone)		

MINERALS

Significant:	Jasper, agate

The area is underlain by gneiss with granodiorite and feldspar porphyry intrusions on the W (harbour) side. Dallasite, porphyry, jasper and agate are reported to occur among the pebble beaches below Dallas Road in Fairfield, Victoria (S of Beacon Hill Park). Dallasite is named after Dallas Road.

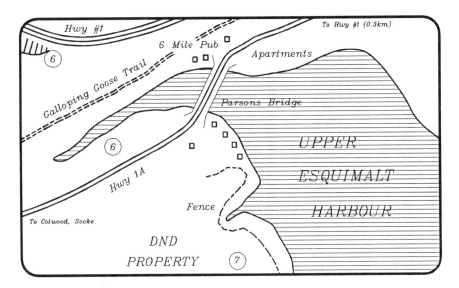

6. MINFILE NUMBER: 092B 020
NAME(S): PARSONS BRIDGE

* Geology

Status:	Past producer, open pit		
NTS Map:	092B 06W	UTM Zone:	10
Latitude:	48 27 13	Northing:	5366590
Longitude:	123 27 26	Easting:	466200
Elevation:	20 m		
Comments:	Limestone outcrop.		
Commodities:	Limestone, iron, copper		

MINERALS

Significant: Calcite, magnetite, chalcopyrite

The Parsons Bridge deposit is located 300 m SW of Parsons Bridge on the W side of Highway 1A at the head of Esquimalt Harbour. A steeply dipping limestone lens, 30 m wide, it is bounded to the E and SE by a granitic intrusive and is intruded by greenstone dykes. The limestone is fine-grained, bluish gray and high in calcium.

Limestone was initially quarried here in 1912 and continued from 1917 to 1922 and 1938 to 1941. Over that time, 7,694 tonnes of limestone were quarried.

Thanks to Mr Gilles Lebrun who reports an extension of the deposit now appears in the new highway cutting (S side).

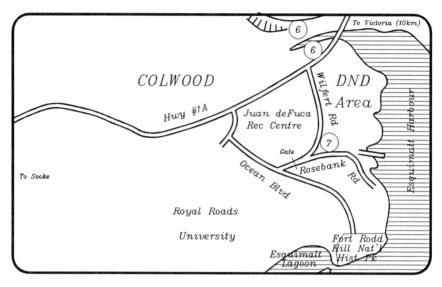

7. MINFILE NUMBER: 092B 024   * Geology
NAME(S): ROSEBANK, DUNNS NOOK

Status:	Past producer, open pit		
NTS Map:	092B 06W	UTM Zone:	10
Latitude:	48 26 37	Northing:	5365500
Longitude:	123 27 29	Easting:	466120
Elevation:	30 m		
Comments:	Quarry is currently within DND fenced area. Limited access.		
Commodities:	Limestone, marble		

MINERALS

Significant:	Calcite
Associated:	Dolomite, chlorite, quartz

The Rosebank deposit is located on the W side of Esquimalt Harbour, 9 km W of Victoria. A band of limestone, 400 m wide, extends W from the shore of Esquimalt Harbour for at least 2 km within greenstone gneiss. It is extensively fractured and one distinct fracture cleavage strikes NW and dips approximately 70 degrees NE. Randomly orientated, greenish, mafic dykes a few centimetres to 15 m wide are locally quite numerous.

The deposit is comprised of very fine-grained, dark bluish-gray to nearly white limestone that commonly displays banding parallel to the fracture cleavage. Most of the limestone has a high calcium composition.

Between 1912 and 1933, over 200,000 tonnes of limestone were quarried. In 1933, the plant was dismantled and the property acquired by the Department of National Defense.

MALAHAT

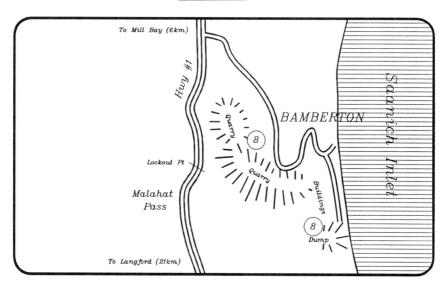

8. **MINFILE NUMBER: 092B 005**   *** Minerals**
       NAME(S): BAMBERTON, ELFORD

Status:	Past producer, open pit		
NTS Map:	092B12E	UTM Zone:	10
Latitude:	48 35 12	Northing:	5381420
Longitude:	123 31 25	Easting:	461390
Elevation:	40 m		
Comments:	Centre of the lower main quarry. Follow the Island Highway # 19 N from Victoria. Turn right after the road descends the Malahat Pass to sea level. Signposted. Currently the site for proposed massive urban development.		
Commodities:	Limestone		

MINERALS

Significant:	Calcite
Associated:	Silica

The Bamberton limestone quarry is located on the W shore of Saanich Inlet, 23 km NW of Victoria.

The deposit is part of a discontinuous carbonate horizon that extends from Cordova Bay NW across Saanich Inlet to the E shore of Shawnigan Lake. The deposit is comprised of limestone lenses. The main lens extends NW from the shore of Saanich Inlet for 700 m and is up to 150 m thick. It is generally dark bluish gray and fine-grained.

The quarries were in operation between 1913 and 1957 while the adjacent cement plant remained in production up to 1980. Between 1913 and 1957, 3.7 million tonnes of limestone were quarried.

SOOKE

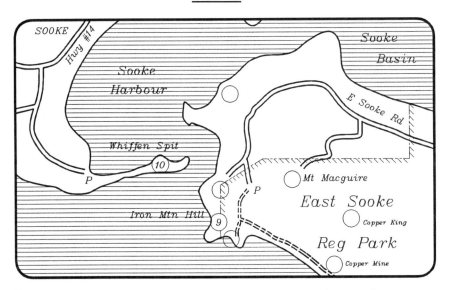

9. MINFILE NUMBER: 092B 007   * **Minerals**
NAME(S): MERRYTH, IRON MOUNTAIN

Status:	Prospect		
NTS Map:	092B 05E	UTM Zone:	10
Latitude:	48 20 27	Northing:	5354200
Longitude:	123 42 30	Easting:	447500
Elevation:	20 m		

Comments: Take the road to the W car park of East Sooke Regional Park. Walk the 1 km to the sea. Up the hill to the right is Iron Mine Hill (092B 065), while further N along the ridge (opposite Simpson Pt) Hill (092B 065) shows similar mineralization. At the coast is Merryth (old mine equipment still visible). In the centre of the Park, high grade anorthosite lenses (plagioclase-rich rock) at Willow Grouse (092B 010) on the NNW flank of Mt Maguire was a past producer of copper, nickel, cobalt, palladium and molybdenum. Be aware that collecting mineral specimens in any park in BC is prohibited.

Commodities: Copper, gold, iron

MINERALS

Significant: Magnetite, chalcopyrite, pyrrhotite, pyrite
Associated: Hornblende

The Merryth deposit is on the SW shore of the Sooke Peninsula, due S of Iron Mine Hill. The main altered zone, made up of hornblende and masses of unaltered gabbro, trends up the hill from the shore at about 25 degrees for 460 m. The deposit is divided into two zones based on a horizontal offset of 15 m to 30 m which is a result of crossfaulting. South of the crossfault the deposit has been called the Merryth zone and to the N it has been called the Iron Mountain zone. The deposit is generally known as the Merryth.

The main mineralization of the Merryth zone is confined within two parallel shears about 30 m apart. The hanging wall contains large amounts of magnetite as replacement grains and as fracture-filling stringers. A small quantity of magnetite occurs in the mineral zone proper. The predominant form of mineralization is the filling of small veinlets and cracks by pyrrhotite, pyrite and late chalcopyrite. The Iron Mountain zone, N of the crossfault, has been examined along its margins and at these points it consists mainly of magnetite and pyrrhotite.

At the E end of the Park, just S of Aylard's Farm, the coast reveals good exposures of sandstone and conglomerate.

| 10. | MINFILE NUMBER: 092B X02 | | ** Gemstones |
| | NAME(S): WHIFFIN SPIT | | |

Status:	Prospect		
NTS Map:	092B 05E	UTM Zone:	10
Latitude:	48 21 20	Northing:	5356400
Longitude:	123 42 35	Easting:	44700
Elevation:	1 m		
Comments:	On the gravel spit at the mouth of Sooke Inlet.		
	WATCH OUT FOR RISING TIDES!		
Commodities:	Flowerstone porphyry, dallasite, agate		

Access the spit off Highway #14 W of Sooke. Winter storms can bring in a variety of beachcombing material, even carnelian and jasper.

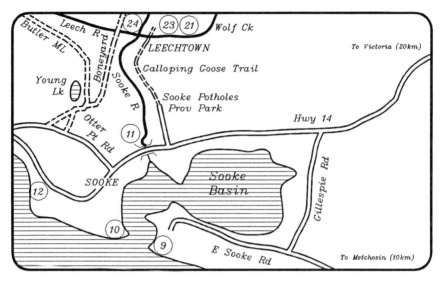

11. MINFILE NUMBER: 092B X05   * **Gemstones**
NAME(S): SOOKE RIVER

Status:	Showing		
NTS Map:	092B 05E	UTM Zone:	10
Lattitude:	48 22 19	Northing:	5360000
Longitude:	123 42 17	Easting:	44800
Elevation:	10 m		

Comments: Placer deposits. For access to the southern end of the river, turn N off Highway #14 at the Sooke River bridge and drive 5 km on the E side of the river to the car park at Sooke Potholes Provincial Park. For the N end, take the Otter Point Rd out of central Sooke, then the Young Lake Rd, and finally the Boneyard Lake Rd which parallels the Sooke River for some distance.

Commodities: Gold

MINERALS

Significant: Magnetite, garnet

Good local gold panning site from Highway #14 to Sooke Potholes Provincial Park; best after winter rains have brought down new colours. Good locations are to be found along the river to N end of Galloping Goose Trail. Green and red garnet (pin-head size) are also common.

12. MINFILE NUMBER: 092B 123   ** Gemstones
NAME(S): SOOKE BAY BEACH

Status:	Showing		
NTS Map:	092B 05W	UTM Zone:	10
Latitude:	48 21 57	Northing:	5357000
Longitude:	123 43 49	Easting:	445900
Elevation:	1 m		
Comments:	Reported to be on the beaches in the Sooke area.		
Commodities:	Gemstones, agate		

MINERALS

Significant:	Jasper, agate, chalcedony
Comments:	Sardonyx is the type of chalcedony found.
Associated:	Calcite

Jasper, agate, sardonyx and calcite crystals are reported to have been found on Sooke beaches. The area is underlain mainly by pillow basalt, breccia and tuff.

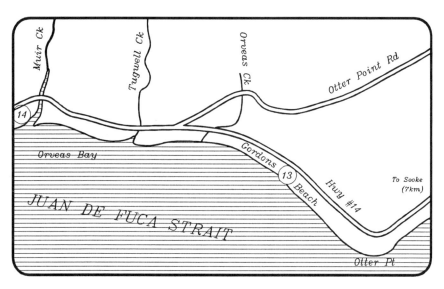

13. MINFILE NUMBER: 092B X03
NAME(S): GORDONS BEACH

** Gemstones

Status:	Showing		
NTS Map:	092B12W	UTM Zone:	10
Latitude:	48 21 50	Northing:	3575000
Longitude:	123 49 35	Easting:	53830
Elevation:	2 m		
Comments:	A few km W of Otter Point.		
Commodities:	Flowerstone porphyry, dallasite, agate		

Beachcombing area, particularly productive after winter storms have re-worked the gravels. At Otter Point, there are outcrops of basalt flows (dark rock) of the Metchosin Formation, with calcite and chlorite infillings.

HIGHWAY #14 TO PORT RENFREW

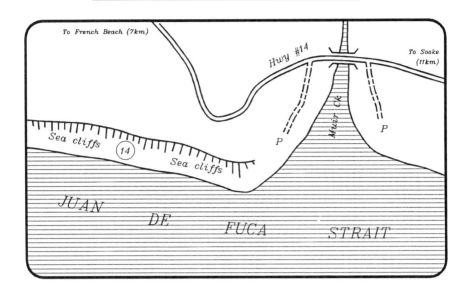

14. MINFILE NUMBER: 092B X04   * Fossils
NAME(S): MUIR CREEK

Status:	Showing		
NTS Map:	092B12W	UTM Zone:	10
Latitude:	48 24 00	Northing:	5358850
Longitude:	123 51 35	Easting:	43540
Elevation:	5 m		
Comments:	About 4.5 km beyond Otter Pt. Park on the W side of the bridge, and walk W for sandstone sea cliffs.		
Commodities:	Fossiliferous sandstone		

About 500 m W of Muir Creek, the low sandstone cliffs are full of clam, snail, mussel and other marine shells. Beware of high tides.

Thanks to Mr Garry McCue for first reporting this site.

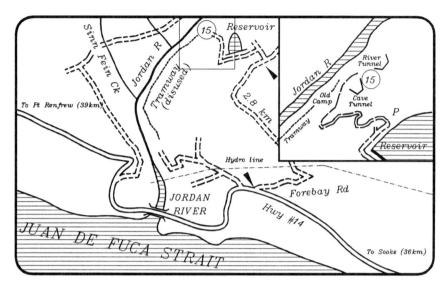

15. MINFILE NUMBER: 092C 073   * **Gemstones**
NAME(S): SUNRO

Status:	Past producer, underground		
NTS Map:	092 C08E	UTM Zone:	10
Latitude:	48 26 55	Northing:	5366450
Longitude:	124 01 54	Easting:	423700
Elevation:	300 m		

Comments: On Jordan River about 3 km from the sea. Spectacular gorge, remote. Portals sealed. Access 2.8 km up Forebay Rd, then L to reservoir, L along the wall, R at end. Track into gorge is 4WD for several km, steep and eroded near the bottom. River Tunnel portal open, but only accessible by experienced climbers. Lots of float in the dumps below portals.

Commodities: Copper, gold, silver, molybdenum

MINERALS

Significant: Chalcopyrite, pyrrhotite, pyrite, molybdenite, pentlandite, copper, cubanite

Comments: Pentlandite occurs locally in pyrrhotite.

Three NW trending bands of gabbro occur, ranging in width from 150 to 900 m, separated by about 1 km of basalt, and known to extend along strike for about 6.5 km. The centre band, from 600 to 900 m wide, is the widest and most important, hosting copper mineralization in shears in basalt along both contacts. The rock is a dark greenish gray coarse-grained hornblende gabbro with conspicuous plagioclase crystals. Some white patches occur in the gabbro where plagioclase has been hydrothermally altered to scapolite. The basalt in the contact zone has a definite hornfels texture.

Microscopic lathes of cubanite have been noted in some specimens of chalcopyrite, and minute blebs and wisps of pentlandite have been seen in pyrrhotite. Much of the pyrite has a striking colloform texture. As many as sixteen mineralized zones have been located on the property since it was discovered in 1915. The zones typically occur in basalt but at least three minor zones are located in areas mapped as gabbro. Production commenced in 1962 and proceeded intermittently for 8 years until 1974.

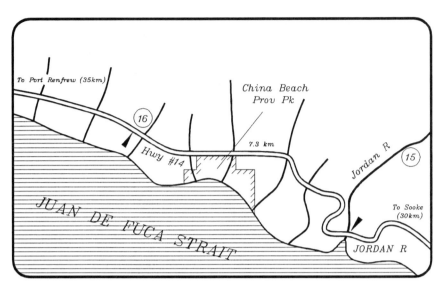

16. MINFILE NUMBER: 092C X05   * Fossils
NAME(S): CHINA BEACH FOSSIL BEDS

Status:	Showing		
NTS Map:	092C 08	UTM Zone:	10
Latitude:	48 26 50	Northing:	5366750
Longitude:	124 08 15	Easting:	416000
Elevation:	125 m		
Comments:	Located on Highway #14 about 7.3 km W of Jordan River bridge. Black cutting on N side of road just after small creek. Boulders below road, formed during construction, offer best material.		
Commodities:	Fossils (bivalves)		

MINERALS

Significant: Thomsonite , analcite, fluorescent calcite

Various bi-valve fossils may be found in the stony conglomerates of the area. Gilles Lebrun reports crystalization adjacent to the fossils reveal microscopic deposits of rare zeolites (analcite) and calcite.

Thanks to Mr Gilles Lebrun for first reporting this site.

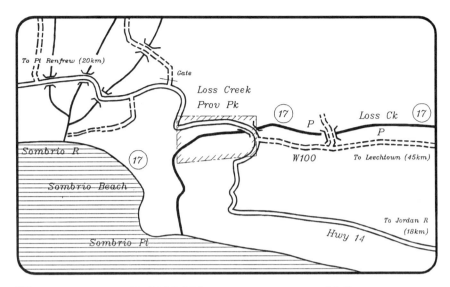

17. MINFILE NUMBER: 092C 044
NAME(S): SOMBRIO PLACERS, LOSS CREEK

* * Gemstones

Status:	Showing		
NTS Map:	092C 08W 092C 09W	UTM Zone:	10
Latitude:	48 29 13	Northing:	5371000
Longitude:	124 17 08	Easting:	405000
Elevation:	10 m		
Comments:	Located on the coast of Vancouver Island between the mouths of the Sombrio River and Loss Creek.		
Commodities:	Gold		

MINERALS

Significant:	Gold

The Sombrio Placers, on the SW coast of Vancouver Island near Sombrio Point, occur in a fairly level coastal area composed of sand, gravel and clay from 60 to 120 m in depth. These gold placers are apparently the remains of a glacial delta deposited by glacial and postglacial rivers that drained SW through the Leech River valley. The E side of the delta is cut by Loss Creek, and the W side, up to 3 km away, by the Sombrio River. The gold is thought to have been derived from quartz veins and stringers known to occur in slate of the Leech River Complex.

The Spaniards first identified gold in the area in 1792; the name "Sombrio" is Spanish for colours. Elaborate camps and engineering works were constructed on the property from 1900 to 1930. Some production was reported to have occurred from 1907 to 1914 using a fifty man monitor and sluice operation. Work continued on the deposit in the 1970s and 1980s.

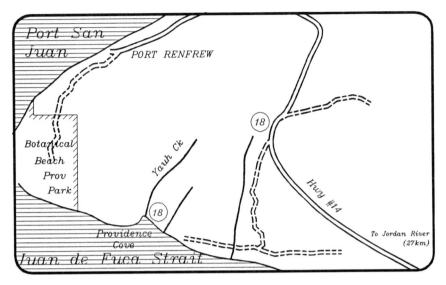

18. MINFILE NUMBER: 092C 071 * **Gemstones**
   NAME(S): SPANISH, PROVIDENCE COVE

Status:	Showing		
NTS Map:	092C 09W	UTM Zone:	10
Latitude:	48 32 24	Northing:	5377000
Longitude:	124 22 06	Easting:	399000
Elevation:	240 m		
Comments:	Centre of major claim group showing attitude of gold bearing-vein.		
Commodities:	Gold		

MINERALS

Significant:	Gold
Associated:	Quartz

A gold nugget was reported to have been found, in 1893, in a small stream flowing into Providence Cove. Further prospecting led to the discovery of several quartz veins, all carrying small quantities of gold on the surface outcrops. Where Highway #14 curves N towards Port Renfrew, a possible source of the gold has been reported; location unspecified.

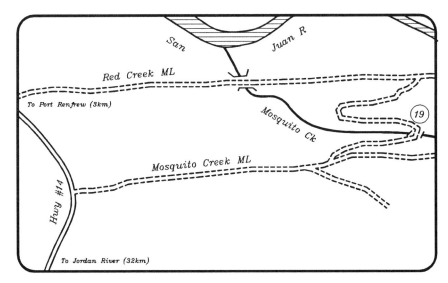

19. MINFILE NUMBER: 092C 059
NAME(S): OX, CAROL 1

* Minerals

Status:	Showing		
NTS Map:	092C 09W	UTM Zone:	10
Latitude:	48 34 01	Northing:	5379900
Longitude:	124 16 51	Easting:	405500
Elevation:	280 m		
Comments:	Near Mosquito Creek, about 6 km from its mouth in the San Juan River.		
Commodities:	Gold		

MINERALS

Significant:	Pyrite, pyrrhotite, arsenopyrite, gold
Associated:	Quartz, carbonate

The quartz veins contain varied amounts of carbonate and tend to occur in areas of abundant siliceous interlayering. A few specimens containing free gold were collected. One sample from a narrow quartz vein assayed 50 grams per tonne gold. Gold in four separate quartz showings within 190 m. One vein is traceable for 46 m. The quartz veins strike approximately SSW. Just to the E, panned concentrate sampling in Mosquito Creek resulted in visible gold grams.

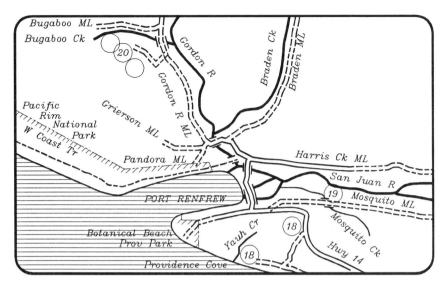

20. MINFILE NUMBER: 092C 027   * Minerals
NAME(S): BADEN POWELL, BUGABOO CREEK

Status:	Showing		
NTS Map:	092C 09W	UTM Zone:	10
Latitude:	48 39 19	Northing:	5390000
Longitude:	124 29 17	Easting:	390400
Elevation:	400 m		
Comments:	There are four other sites in the upper Bugaboo Creek area (MINFile # 022025) with similar showings.		
Commodities:	Iron, magnetite		

MINERALS

Significant:	Magnetite, pyrite
Alteration:	Magnetite, limestone

These ore deposits, all centred around the W side of Gordon River and the S side of Bugaboo Creek, consist of magnetite occurring within zones of pyroxene skarn formed along the jointing of diorite and a limestone roof of similar age. Actinolite is a minor constituent in the zone of alteration.

The magnetite occurs as large, irregular massive bodies entirely surrounded by skarn. It is essentially free of impurities and has only a small percentage of included sulfides. Assays of magnetite yielded grades up to 69% iron with 0.5% sulphur. The only sulfides present are pyrite and pyrrhotite.

Reports describe the Conqueror claim at 480 m altitude, "where the creek intersects a body of magnetite about 15 m high, over which the creek forms a waterfall." The deposit lies within crystalline limestone (almost white marble) and diorite. Maximum width of the ore body in the creek is 30 m.

In one report, magnetite was traced for over 100 m on the flank of a ridge, with a 40 m tunnel. Elsewhere, "magnetite outcrops along the face and brow of the hill for 30 m. About 15 m below the top of the hill, a 30 m tunnel has been driven."

The new road up Gordon River cuts white marble and crystalline limestone at several locations.

LEECHTOWN AREA

Best access into this area is from Sooke. Turn N on the Otter Point Rd, past the golf course, and at 5 km turn R onto the Butler Mainline (Young Lake, Tugwell Lake). Less than 1 km along this road, pass Camp Barnard on Young Lake and the blacktop ends. A km along the Butler Mainline, the Boneyard Lake mainline takes off right. Both are slippery in winter and dusty in summer. The Boneyard gives access to the historically interesting geological area of Leechtown (sites 21-24), where gold was first discovered in 1864 by Lt. Leech of the Royal Navy.

The Leech River Fault is a major E-W line (now occupied by Loss Creek, Old Wolf Creek, and Goldstream River) that separates the younger seafloor volcanics to the S from the much older Pacific Rim Terraine to the N.

Placer deposits were discovered in the 1860s and at that time were extensively worked. It is believed that over three thousand men were engaged in placer mining at one time along Leech River. By 1876, it was estimated that $100,000 worth of gold had been recovered. Later estimates place the actual value between $100,000 and $200,000. Signs of old workings are seen along the river upstream from the Sooke River, a distance of about 6.5 km to a point 1.5 km beyond the first fork.

According to one account, the run of gold turned up the North Fork but rapidly diminished and ran out above the falls in the Devil's Grip. Between 1924 and 1945 a recorded 192 ounces of gold were recovered. Of the tributaries to Leech River, Martin's Gulch is notable for the gold that was found for a distance of 2 km up from Leech River.

It appears that most of the gold was derived from bars or in crevices in the bedrock of the river bed, or from benches along the side of the river. The gold recovered from the benches was mined either at a depth of 3 to 5 m and 3 m above river level on a clay "false bedrock" of a low bench on the N side of the Leech River that extends 400 m upstream from its junction with the Sooke River or on the bedrock beneath the shallow overburden on a rock bench about 3 m above river level that extends more or less continuously on one side of the river or the other, at least as far as the first fork in the river about 5 km upstream from Sooke River. Nuggets varying in size from 15.6 to 31.1 gms have been recovered.

Access to the more westerly areas (sites 25-28) is best over the Butler Mainline from Otter Point Rd.

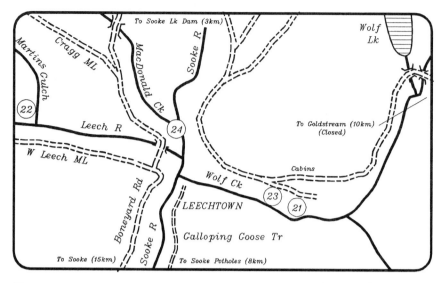

21. MINFILE NUMBER: 092B 012   *** Minerals**
NAME(S): EAGLE

Status:	Past producer, underground		
NTS Map:	092B 05E	UTM Zone:	10
Latitude:	48 29 24	Northing:	5370766
Longitude:	123 40 36	Easting:	450002
Elevation:	267 m		
Comments:	On Wolf Creek, 2.5 km E of the confluence with the Sooke River.		
Commodities:	Talc		

MINERALS

Significant:	Talc
Associated:	Magnetite, dolomite, calcite

The Eagle deposit is on Wolf Creek, E of the confluence with the Sooke River. In 1923, 250 tons of talc was mined and shipped to a plant in Sidney, British Columbia. The Eagle, as well as three other talc showings, lie on the Leech River fault. Near the workings, the slates are black, carbonaceous and severely crushed and folded. Talc occurs in three narrow, lens-shaped bodies paralleling the schistosity in the slates. A 2.1 m thick body outcrops in the top bank and expands to 4.5 m thickness 12 m below. Some 15 m below the first showing, a 3-3.5 m talc lens is enclosed in talcose slates and black, soft, slaty argillite. A third 2.1 m thick talc body is found exposed in the creek with another talc outcrop appearing on strike, 1.5 m W.

The lenses are homogeneous and mottled gray with faint black specks (magnetite?). The talc is light greenish gray, granular, very friable and crushes to an off-white powder. The crude ore is 50% talc and 38% dolomite and calcite.

22. MINFILE NUMBER: 092B 078   *** Gemstones
NAME(S): LEECH RIVER PLACER, MARTIN'S GULCH, KENNEDY FLAT

Status:	Past producer, open pit		
NTS Map:	092B12W, 092B12E, 092B05E	UTM Zone:	10
Latitude:	48 30 02	Northing:	5372000
Longitude:	123 44 40	Easting:	445000
Elevation:	200 m		
Comments:	The Leech River placers occur along Leech River (and tributaries) between its confluence with the Sooke River and a point upstream about 6.5 km.		
Commodities:	Gold		

MINERALS

Significant:	Gold

Placer gold occurs almost exclusively in the gravels of the streams that drain the area underlain by the slaty schists of the Leech River Complex. Fairly coarse gold may be found in the gravels of virtually all these streams. The gold in the recent gravel deposits has likely been derived from the numerous quartz veins that occur in the slaty schists. These veins are seldom more than small stringers and lenses a few centimetres wide and a metre or so in length. The only metallic minerals in the veins are a little pyrite or chalcopyrite and free gold. The veins are generally too small and too barren to be profitably mined.

23. MINFILE NUMBER: 092B 050   *** Gemstones
NAME(S): WOLF CREEK

Status:	Past producer, open pit		
NTS Map:	092B 05E	UTM Zone:	10
Latitude:	48 29 18	Northing:	5370600
Longitude:	123 41 05	Easting:	449400
Elevation:	220 m		
Comments:	On Old Wolf Creek, about 2 km upstream from Leechtown		
Commodities:	Gold		

MINERALS

Significant:	Gold

Old Wolf Creek flows W along the Leech River fault which separates the Leech River Complex slates and schists on the N, from Metchosin Volcanics on the S. The topography shows the stream to have cut down through the bedrock leaving a series of gravel covered benches. Placer gold, found in the gravel, is believed to have been derived from small, but numerous, gold-bearing quartz stringers hosted by the Leech River rocks.

The creek was worked in the early 1930's at a location about 2 km above Leechtown where, about 9 m above the creek, an old creek channel was found.

24. MINFILE NUMBER: 092B 051   ** Minerals,
NAME(S): EASTERN STAR   Gemstones

Status:	Showing	Mining Division:	Victoria
NTS Map:	092B 05E	UTM Zone:	10
Latitude:	48 29 45	Northing:	5371436
Longitude:	123 42 36	Easting:	447545
Elevation:	160 m		
Comments:	At the junction of McDonald Creek and Sooke River, about a km N of old Leechtown.		
Commodities:	Talc, gold		

MINERALS

Significant:	Talc, gold
Alteration:	Talc

A number of talc showings occur on or adjacent to the Leech River fault. The talc occurs in shear zones within slates and schists of the Leech River Complex. The showings are also of interest because of the native gold which is sometimes found in the gouge material lying between the talc and the slate country rock. On the Eastern Star showing, a body of talc 2.3 m long was exposed in an opencut.

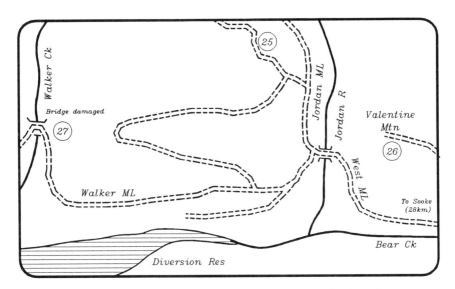

25. MINFILE NUMBER: 092B 109   * Minerals
NAME(S): JORDAN RIVER

Status:	Showing		
NTS Map:	092B 12W	UTM Zone:	10
Latitude:	48 30 30	Northing:	5373000
Longitude:	123 56 03	Easting:	431000
Elevation:	500 m		
Comments:	An interesting area, but specimens are small and hard to find.		
Commodities:	Sillimanite, kyanite, mica schist		

MINERALS

Significant:	Sillimanite, staurolite, garnet, kyanite, andalusite, tourmaline

In a few places, notably in the vicinity of Jordon River, the quartz-biotite and quartz-sericite schists contain large amounts of garnet, staurolite and sillimanite. To the W, kyanite has been observed. The exposed S third to half of the Leech River unit contains staurolite and andalusite, with the latter forming porphyroblasts up to 20 cm in length.

The sillimanite occurs in short to long slender prisms, of rectangular or hexagonal cross-section, that are rudely oriented parallel to foliation. They have a maximum length of about 5 cm and a maximum width of 4 cm or 5 cm. Garnet occurs in small (averaging less than 2 mm in diameter) pinkish, dodecahedral crystals. Staurolite, not usually conspicuous in hand specimens, forms small, yellowish grains between the lamellae of biotite and quartz; in a few places distinct single and twin crystals are seen.

Within a relatively narrow zone W of the Jordan River are carbonaceous andalusite-staurolite-biotite rocks. These rocks contain euhedral twinned staurolite crystals up to 3 cm long, garnet averaging 1 cm across, and erratic black tourmaline. Most of the large andalusite crystals have been altered to chlorite, biotite and sericite.

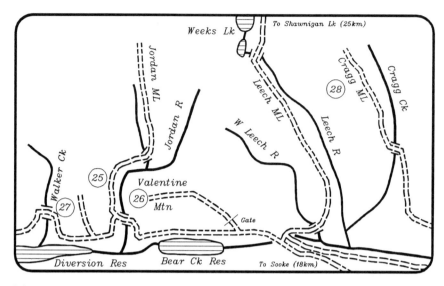

26. MINFILE NUMBER: 092B 108   ** Minerals
NAME(S): VALENTINE MOUNTAIN

Status:	Developed Prospect		
NTS Map:	092B12W	UTM Zone:	10
Latitude:	48 31 04	Northing:	5374000
Longitude:	123 52 58	Easting:	434800
Elevation:	820 m		
Comments:	Located about 1.5 km N of the E end of the Bear Creek. Reservoir. Gate locked.		
Commodities:	Gold, copper		

MINERALS

Significant:	Pyrite, arsenopyrite, pyrrhotite, chalcopyrite
Associated:	Quartz

The Valentine Mountain occurrence lies within the Leech River, N of the Leech River fault. Narrow quartz veins cutting both meta-sedimentary and meta-volcanic rocks carry spectacular coarse free gold. These veins are from 2 cm to 50 cm in width, strike about 67 degrees and are nearly vertical in dip. The veins seldom exceed 10 cm in width and can be traced for ten's of metres, apparently barren for parts of their length. The zone, along which these gold-bearing veins occur, trends E for a distance of almost 3 km and is from 200 m to 300 m in width.

Rare sulphide minerals generally consist of disseminations of pyrite, arsenopyrite, pyrrhotite and occasionally chalcopyrite. Large arsenopyrite crystals have locally been fractured and infilled by fine gold. Gold smears have also been noticed along fracture surfaces and as small specks in the wallrock, a few centimetres from vein material. Most of the higher grade gold values appear in either fracture or quartz veins within biotite schist.

27. MINFILE NUMBER: 092B 111   *** Minerals
NAME(S): PEG

Status:	Showing		
NTS Map:	092B12W	UTM Zone:	10
Latitude:	48 30 38	Northing:	5373300
Longitude:	123 59 18	Easting:	427000
Elevation:	700 m		
Comments:	Centre of mineralized area, with likely more potential. This is the only pegmatite deposit identified to date on the Island. Pegmatite fields in Ontario, Brazil and elsewhere are the source of many spectacular gemstones.		
Commodities:	Copper, beryl, feldspar, gemstones		

MINERALS

Significant:	Pyrite, chalcopyrite, bornite, beryl, tourmaline, feldspar
Associated:	Feldspar, rhodochrosite, apatite, spinel

Scattered, sparse sulphide mineralization found in the area consists of pyrite, chalcopyrite and bornite. There is reported to be good correlation between the occurrence of rhodochrosite and copper mineralization. Beryl, tourmaline, apatite, spinel and feldspar are reported to occur in two pegmatite veins in the area.

Access is from the bridge spanning the Jordan River between Diversion Reservoir and Bear Creek Reservoir. Instead of heading N on Jordan Mainline, turn S then W on Walker Mainline. After 4 km, track deteriorates and turns N to damaged bridge over Walker Creek. Pegmatites have been found in the road banks next to the bridge, and NW above.

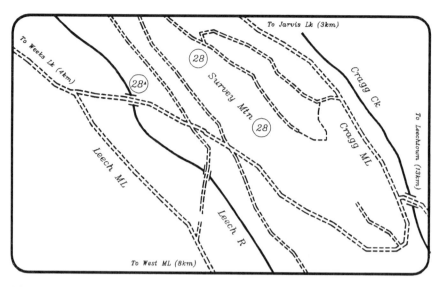

28. MINFILE NUMBER: 092B 147
NAME(S): SURVEY MOUNTAIN

* Minerals

Status:	Showing		
NTS Map:	092B12W	UTM Zone:	10
Latitude:	48 33 41	Northing:	5378800
Longitude:	123 47 46	Easting:	441250
Elevation:	900 m		
Comments:	Near the summit of Survey Mountain.		
Commodities:	Copper, zinc		

MINERALS

Significant:	Pyrite, pyrrhotite, chalcopyrite, sphalerite
Associated:	Quartz, calcite

An extensive gossan zone occurs along the crest of the Survey Mountain ridge. The zone is generally less than 4 m wide and may extend for up to 1 km in length. The gossan zone is hosted in pillowed flows and breccias. Sulfides include pyrite, pyrrhotite and trace amounts of chalcopyrite and possibly sphalerite.

Massive sulfides were observed in large irregular blocks near the peak of Survey Mountain, near the southern outcrop of the gossan zone. They were found to contain up to 60% pyrite and pyrrhotite which occurs in irregular veins and veinlets. Gangue minerals include quartz and possibly calcite and chlorite.

There are unconfirmed reports that significant quartz crystals are to be found in Crystal Creek, on the W side of the mountain. Location uncertain. Where the Leech River is crossed by the logging road (site 28[A]), gold is reported in the stream. Currently under claim.

GULF ISLANDS

Most of the Gulf Islands are Nanaimo Group sedimentaries. A major fault line, possibly an extension of the Lake Cowichan Fault, cuts through Salt Spring Island, with the southern third (S of a line between Burgoyne Bay and Fulford Harbour) being made up of the much older (and more interesting) Mt Sicker Group.

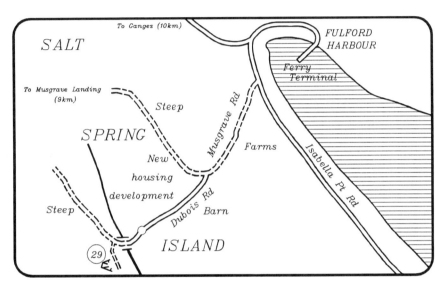

29. MINFILE NUMBER: 092B 074
NAME(S): HOLLINGS

*** **Gemstones**

Status:	Producer, open pit	Mining Division:	Victoria
NTS Map:	092B11W	UTM Zone:	10
Latitude:	48 44 53	Northing:	5399336
Longitude:	123 28 10	Easting:	465490
Elevation:	300 m		
Comments:	On southern Salt Spring Island. Take the Musgrave Landing road, turn L onto Dubois (blacktop) and follow to its end. Old material everywhere. Latest report has the quarry being used as a building site for a house.		
Commodities:	Rhodonite, gemstones		

MINERALS

Significant:	Rhodonite
Associated:	Quartz, jasper, spessartite, rhodochrosite
Alteration:	Garnet, magnetite

The Hollings rhodonite deposit consists of a lens, convex at top and bottom, reaching a maximum thickness of 5 m and a minimum length of 30 m. The banded variety contains abundant quartz, jasper and minor amounts of spessartite garnet. Rare fragments of green chert host rock are present in the rhodonite. Rhodochrosite surrounds calcite veins and bands. There are numerous quartz, minor rhodonite and minor neotocite veins cutting the rhodonite.

The lower 25 cm of the rhodonite lens consist of interbeds of rhodonite and calcareous magnetite-garnet schist, followed by a 9.5 m sequence of argillaceous metachert, cherty crystal tuff with idiomorphic garnet and minor recrystallized limestone. Directly up-section from the rhodonite lens is a 0.5 m band of argillaceous garnet-magnetite metachert. Still further up-section, an igneous intrusion (unspecified type) has obliterated the section. Glacial drift and alluvium cover the sequence further down.

The rhodonite is considered to be of excellent gem quality and the deposit has been quarried commercially on a small scale. The material is sold in chunks of various sizes, by the pound or in sawed blocks and slabs. Most of the production has been sold in rock shops for amateur use. The property was sold out of the Hollings family in 1992. The quarry is currently close to the scene of a housing development. Be sensitive to trespass violations.

A second (undeveloped) rhodonite outcrop is reported at PATRICA (092B 073) less than 1 km to SE.

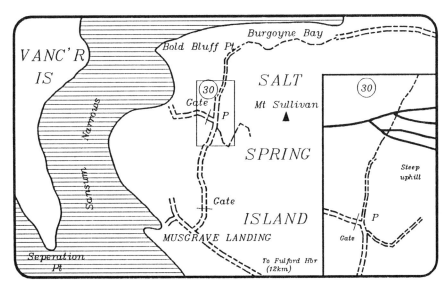

30. MINFILE NUMBER: 092B 030
NAME(S): MESABI

** Gemstones, minerals

Status:	Showing	Mining Division:	Victoria
NTS Map:	092B13E	UTM Zone:	10
Latitude:	48 46 25	Northing:	5402200
Longitude:	123 32 15	Easting:	460500
Elevation:	300 m		

Comments: On the W slope of Mount Sullivan on Salt Spring Island. Drive to Musgrave Landing. 600 m before dock, turn R up a track, passing a gate. After 2 km, pass a locked green gate on left. Proceed straight ahead (N) to washouts in old road (600 m). Turn NW downhill on steep ground for about 200 m.

Commodities: Iron, magnetite

MINERALS

Significant: Magnetite, hematite, jasper
Associated: Quartz
Alteration: Hematite

The mineralized zone strikes intermittently down and across a hillside for about 150 m. It consists of lenticular bands of blood-red jasper, interlayered with schist containing streaks, bands and lenses of magnetite and smaller amounts of hematite. Some of the ore masses are up to 45 m long and 3 to 6 m wide. The deposit is cut by irregular veins and stringers of white quartz. Two other showings are recorded by following the closest stream up Mt Sullivan to the upper valley to chert, quartz and jasper.

There is an unconfirmed report that when driving from Fulford Harbour to Musgrave Landing, there is a quarry on the right side (E) containing fossil deposits, found 1.4 km after the Mt Tuam Road turns to the left.

COBBLE HILL

The Cobble Hill mineral sites lie within a triangle formed by three significant faults. To the S, an E-W fault defines the San Juan and Koksilah River valleys; to the N, a NW-SE trending fault runs through Cowichan lake and Cowichan Bay; to the W, a minor SE-NE trending fault cuts the coast just S of Chemainus.

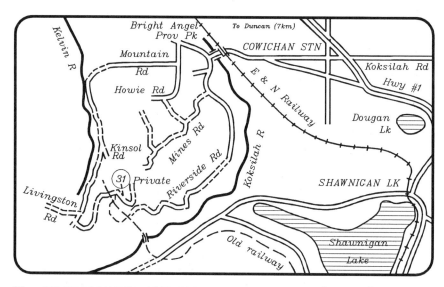

31. MINFILE NUMBER: 092B 035   * Minerals
NAME(S): VIVA, KING SOLOMON

Status:	Past producer, underground		
NTS Map:	092B12E	UTM Zone:	10
Latitude:	48 40 55	Northing:	5392100
Longitude:	123 41 44	Easting:	448800
Elevation:	320 m		
Comments:	The tailings dump and surrounds have shown some interesting mineralization, and are always worth re-working.		
Commodities:	Copper, silver		

MINERALS
Significant: Chalcopyrite, pyrrhotite, pyrite

The Viva is a skarn deposit consisting of pods of pyrrhotite, pyrite, magnetite and chalcopyrite occurring along fractures within chert. The chert is also cut by epidote-filled fractures.

Access is from Cowichan Station (follow signs off Island Highway to Bright Angel Park). After crossing under the railroad and over the river, take Howie and then Mountain Rds.

Just 50 m after road crosses old railway line and climbs E up hill, a track climbs N to King Solomon mine shaft. Other shafts (on private land) follow this dyke up the hill. Skarn, chert, marble, zinc, copper, magnetite and garnet are reported.

Scenic route is to turn left onto Riverside Rd. After 5 km the black top ends; another 3 km brings you to campsite. Access through campgrounds.

MOUNT SICKER

Massive sulfides were discovered on Mount Sicker in the late 1800s and production issued from three separate underground mines (Lenora – 092B 001, Tyee – 092B 002 and Richard III – 092B 003) for several years. These mines were later held as one operating mine, the Twin J mine (1942–1952). The rocks in the mine, and nearby, include cherty tuffs, graphitic schists, rhyolite porphyry and diorite. Two types of ore are found in association with cherty tuffs and graphitic schists: a barite ore consisting of a fine-grained mixture of pyrite, chalcopyrite, sphalerite and a little galena in a gangue of barite, quartz and calcite; and a quartz ore consisting mainly of quartz and chalcopyrite.

Road access is best only as far as Lenora, where a huge dump is currently used as a playground for mountain and scramble bikers. Thereafter, the roads are very steep and heavily washed out. The whole area offers excellent prospecting, but best to start at the mine dumps. Beware of old shafts; many have lost their covers, and present serious obstacles to the unwary.

The Sicker Group is the oldest formation found on Vancouver Island, and offers some of the most interesting geology. The rhodonite deposits of Salt Spring Island and Hill 60, the jasper deposits of the Upper Nanaimo basin, and the copper deposits of Mount Sicker and Myra Falls are all found within its horizons.

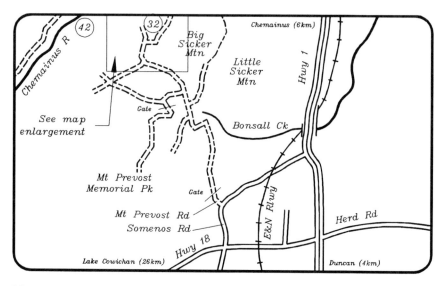

32. MINFILE NUMBER: 092B 001   ** Minerals
NAME(S): LENORA, TYEE, RICHARD III

Status:	Past producers, underground		
NTS Map:	092B13W	UTM Zone:	10
Latitude:	48 52 03	Northing:	5412800
Longitude:	123 47 17	Easting:	442200
Elevation:	350 m		
Comments:	Centre of a number of turn-of-the-century mineral operations. Access is off the Cowichan Lake road. Follow signs to Mt. Prevost Memorial Park until on top of the mountain, then bear NW. A rough road brings you out at the Tyee tailings dump. Beyond this point, the tracks are barely walkable; 4WD is of little use to reach Lenora.		
Commodities:	Copper, gold, silver, lead, zinc, cadmium, barite		

MINERALS

Significant:	Chalcopyrite, sphalerite, galena, barite, pyrite
Associated:	Barite, quartz, calcite

The two main ore bodies, known as the N ore body and the S ore body, are long, lenticular deposits lying along two main dragfolds in the band of sediments. The N ore body measures about 500 m along strike, while the S ore body, which is 45 m from the N, measures 640 m along the strike. The property has undergone steady exploration by various companies from 1964 to present.

In the late 1890s, two competing owners built two mines, towns, churches, railways and smelters, where common sense would have deemed one of each enough. This bitter rivalry resulted in both operators going broke in 1907, and the mines were sold to a British consortium, which combined the two operations and continued until 1921.

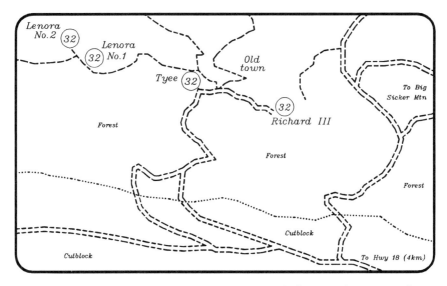

In its day, the pinion railway track up Mount Sicker was the steepest in Canada, and included hair-raising bends and no less than three switchbacks. Locomotive accidents were frequent. Nevertheless, dashing Victorians considered it an exciting experience to ride the rails up and down Mount Sicker on weekends.

COWICHAN VALLEY

A major E-W trending fault cuts along the N shore of Cowichan Lake, bisecting the area geologically, being the contact line where the basalts and pillow lavas of the S meet the Sicker volcanics to the N.

To the S of the lake the primary showings are copper sulfides and quartz veins, while to the N are numerous outcroppings and showings of red and pink jasper, chert and rhodonite. These popular collecting materials stretch N as far as Horne Lake, where dallasite is also found, and provide an excellent hunting ground for semi-precious minerals.

To reach Cowichan Lake, turn off the Island Highway #1 at 7 km N of Duncan on the paved road posted to Youbou and Bamfield.

To get to the Chemainus River area (N of, and parallel to, Lake Cowichan), it is best to turn W off the Island Highway 0.5 km N of the Chemainus traffic lights. This takes you first past a large mill, after which a good tar and then gravel road 30 km long runs up the Chemainus River. Beware of logging trucks. At the time of this writing, the road W of the Rheinhart Lake turnoff has been made impassable.

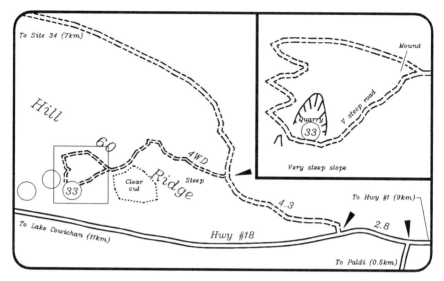

33. MINFILE NUMBER: 092B 027
   NAME(S): HILL 60

*** Gemstones, minerals

Status:	Past producer, open pit		
NTS Map:	092B13W	UTM Zone:	10
Latitude:	48 49 32	Northing:	5408305
Longitude:	123 58 35	Easting:	428334
Elevation:	800 m		
Comments:	Located on the very steep S side of the Hill 60 Ridge near its E end, overlooking the Cowichan Valley. Stunning location. Access 12.8 km from Highway #1 turnoff, or 2.8 km after Paldi turnoff. After 4.3 km on old logging road, a very steep, very rutted road leads to top of E end of Hill 60 Ridge. Close to the quarry, the road forks. The upper road, although less steep, is closed. The lower road is suitable only for tracked vehicles. Use extreme caution.		
Commodities:	Manganese, rhodonite, gemstones, copper		

MINERALS

Significant:	Rhodonite, jasper, chalcopyrite, bornite
Associated:	Quartz, calcite, rhodochrosite, spessartine

The famous Hill 60 manganese deposit, mined by open pit in 1919 and 1920, is underlain by tuffaceous cherts. The rocks associated with the manganese occurrence are thinly banded, green, cream and red tuffaceous cherts locally containing lenses of massive red jasper. These rocks are cut by a few mafic dykes near the Hill 60 workings.

The manganese ore outcropped for a distance of 33 m at a strike of 80 degrees and a dip of about 70 degrees to the SE, along the crest of the hill. The ore consisted mainly of a mixture of hard, compact oxides of manganese (pyrolusite), grading from highly siliceous material along the walls (rhodonite) to a relatively pure oxide at the centre of the ore body. The central portion of highly oxidized ore was about 6.5 m across and 4.6 m in depth. An average sample of the deposit was reported to contain 37% manganese. The rhodonite varies in colour from pink to watermelon red. It is predominantly massive with minor irregular-shaped masses of quartz and the yellow manganese garnet, spessartite.

Toward the periphery of the deposit, these three minerals occur in parallel bands, with quartz predominant. Rare fragments of green chert occur in the rhodonite. Chalcopyrite and bornite occur as disseminations in the rhodonite and jasper. Numerous veins of quartz and fracture-fillings of paler pink rhodonite cut the rhodonite lens. Fault gouge occurs along the contact between the rhodonite and the country rock. Thin section and x-ray diffraction analysis confirm the presence of calcite, rhodochrosite, quartz and rhodonite in the gouge. Thin lenses of rhodonite are present in the tuffaceous cherts approximately along strike from this deposit but not continuous with it.

The deposit was discovered in 1918, and in 1919 and 1920, some one thousand tonnes of ore was shipped. Around 1985, an unknown quantity was mined for jewellery and carvings. Early operators used an aerial tramway from the mine, down the steep S face of Hill 60 to the railway below, thus avoiding the very steep access roads.

From the Island Highway, take the Lake Cowichan road. At the 12.8 km mark, turn right onto a rough track that climbs left upslope. At 4.3 km, turn left up 4WD track. Very steep. About 3 km brings you to a clear cut and then the open ridge. Proceed over the ridge onto S side for 1 km to where track forks. Either track leads to cutting. The left track is more direct. Other manganese deposits occur E and W along the slope.

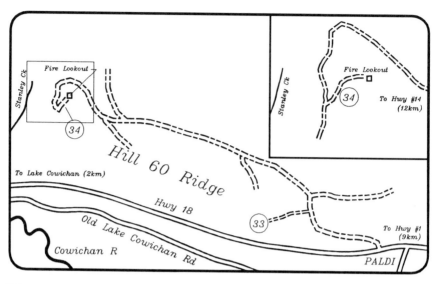

34. MINFILE NUMBER: 092C 116   ** Gemstones
NAME(S): STANLEY CREEK, COW 7

Status:	Showing		
NTS Map:	092C16E	UTM Zone:	10
Latitude:	48 51 29	Northing:	5411950
Longitude:	124 00 56	Easting:	425500
Elevation:	850 m		
Comments:	Rhodonite showing located on the Cow 7 claim, E of the head of Stanley Creek. Snow often stays until June.		
Commodities:	Gold, silver, copper, manganese, rhodonite		

MINERALS

Significant:	Pyrite, chalcopyrite, magnetite, rhodonite
Associated:	Quartz, jasper
Alteration:	Hematite limonite

The Stanley Creek showing is located 4 km NE of Lake Cowichan, E of the head of Stanley creek. The rhodonite has been known of since about 1939 and exploration in 1986 discovered sulphide mineralization in the area.

Rhodonite occurs in thinly laminated chert and cherty tuff of the Fourth Lake Formation. The showing consists of two irregular lenses of rhodonite, parallel to the bedding, about 5 cm to 30 cm wide and 6 m long.

A hematitic chert (iron formation) horizon has been traced for 700 m, possibly extending along strike for several kilometres. The horizon is up to 10 m wide and hosts pyrite and magnetite. Several fault zones cut this unit and, where exposed, are enriched in manganese, barium, zinc and anomalous gold values. These may be the source of the well mineralized float found on the property. The bed, composed of blue-gray cryptocrystalline quartz (sporadically jasperoidal), contains up to 5 per cent pyrite and specular hematite and several per cent magnetite.

Best access is the same route as for Hill 60 (092B 027). About 12.8 km W along Cowichan Lake road from Island Highway, turn N onto track. Follow for about 13 km (4WD near end) as it rises along the N side of Hill 60. At the very NW end of the Hill 60 Ridge the track breaks out onto flat country. Shortly before reaching the headwaters of Stanley Creek, the track turns due S, then after 900 m hairpins left (NE) up to fire lookout. Some 250 m after hairpin, quarry is S of road.

Numerous other manganese ore, blue-gray marble and flowerstone porphyry outcrops reported in the local area.

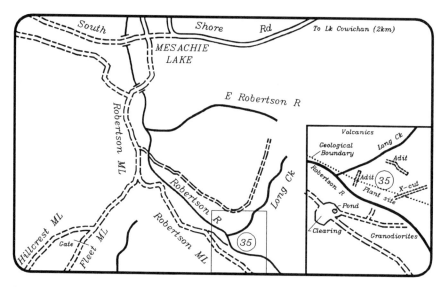

35. MINFILE NUMBER: 092C 039
NAME(S): ALPHA-BETA

* Minerals

Status:	Past producer, underground		
NTS Map:	092C 09E	UTM Zone:	10
Latitude:	48 44 01	Northing:	5398200
Longitude:	124 05 24	Easting:	419850
Elevation:	350 m		
Comments:	Approximately 10 km S on Robertson ML from Mesachie Lake crossroads. Located just E to SE of the junction of Long Creek and Robertson River. Dump prospecting.		
Commodities:	Copper, silver, gold, iron		

MINERALS

Significant:	Chalcopyrite, magnetite, pyrite
Alteration:	Garnet, epidote

The original showings were located in 1904 at the confluence of the Robertson River and "Long" Creek. Alpha-Beta was acquired in the early 1960s by Alberta Mines Limited and developed. The host skarn is known to attain widths in excess of 27 m. A high grade series of copper ores have been found. By November 1963, shipping-grade ore had been depleted and the mining operations were terminated.

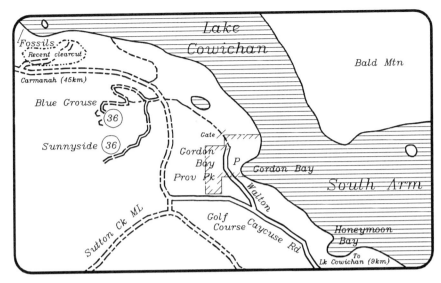

36. MINFILE NUMBER: 092C 108 (also 017)   * * **Minerals**
NAME(S): SUNNYSIDE, BLUE GROUSE

Status:	Past producer, underground		
NTS Map:	092C16E	UTM Zone:	10
Latitude:	48 50 15	Northing:	5409900
Longitude:	124 13 23	Easting:	410250
Elevation:	330 m		

Comments: Originally accessed through Gordon Bay Provincial Park. Now close to the new Carmanah ML. Blue Grouse offers easy access via road to the base of tailings. View from mine portals impressive. One portal open, due to erosion.

Commodities: Copper, silver

MINERALS

Significant: Chalcopyrite, pyrrhotite, pyrite, arsenopyrite, cuprite, calcite
Associated: Quartz, zeolite, malachite
Alteration: Garnet, epidote, actinolite, zeolite

The Sunnyside deposit was part of the Blue Grouse mine (092C 017) which is located on the S side of Cowichan Lake, 5 km NE of Honeymoon Bay. The Sunnyside workings are about 800 m S of the main Blue Grouse workings. Developmental work on the Sunnyside deposit was first reported in 1906. The mine was abandoned in 1960 with some reserves left at the Blue Grouse main workings.

Lenses of chalcopyrite occur in a quartz gangue along the contact zone which is up to 100 m wide. Garnet-epidote-actinolite skarns are also developed in limey tuff, limey sediments and limestone, apparently interbedded with the upper portions of basalts.

Located just W of Gordon Bay Provincial Park. Access is via Highway 18 to Honeymoon Bay, then along gravel road toward Caycuse on S side of

Lake Cowichan. A bike track goes through the park for about 1.5 km. Normal access is to stay on Caycuse road, which leads into Carmanah ML. From Honeymoon Bay Fire Station, drive 5.3 km to turnoff W uphill onto old mine road. Track passes mill and crushers.

The shoreline W of a small island on the S side of Cowichan Lake is known for its fossil deposits. Best approach by boat, although recent logging between the Carmanah ML and the shore allows bushwacking access.

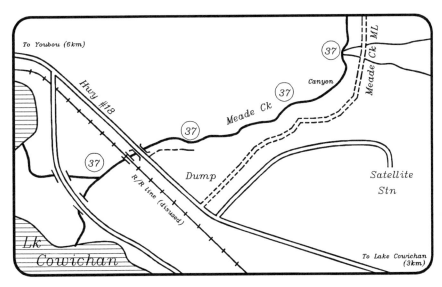

37. MINFILE NUMBER: 092C 057
NAME(S): MEADE CREEK

** Gemstones

Status:	Past producer, open pit		
NTS Map:	092C 16E	UTM Zone:	10
Latitude:	48 50 45	Northing:	5410650
Longitude:	124 04 44	Easting:	420850
Elevation:	260 m		
Comments:	Approximate location at centre of canyon. There may be a staking reserve over part of this area. Check before panning.		
Commodities:	Gold		

MINERALS
Significant: Gold

The Meade Creek placer is 4 km W of Lake Cowichan village. The leases extend upstream from about 150 m above the now-disused CNR railway bridge, covering more than 1.5 km of the creek bed.

The panning and sluicing was done along a 700 m stretch, between 0.8 km and 1.5 km above the bridge. The creek flows through a canyon, which contains stream debris ranging in size from fine sand to boulders. In the canyon, fine colours are seen in most pans containing bedrock material and in sand among the roots of the trees near the high-water mark. Gold is also reported to have been panned from overburden near the creek as much as 6 m above high water level outside the canyon.

Upstream lenses of rhodonite and manganiferous garnet in red cherty tuffs (092C 115) wash downstream, providing interesting material.

Access is 600 m E of Meade Creek bridge. Drive N past incinerator. Bottom of canyon offers deep pools and moss covered boulders. At top end, easy access is possible where two wooden bridges span tributaries.

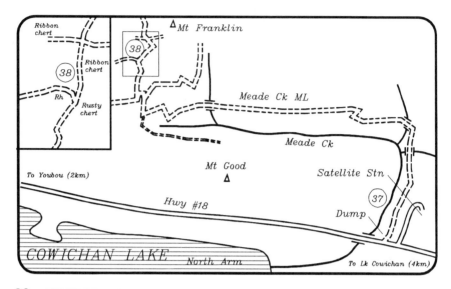

38. MINFILE NUMBER: 092C 113   * Gemstones
NAME(S): ROCKY, OSIRUS A, MT FRANKLIN

Status:	Past producer, open pit		
NTS Map:	092C 16E	UTM Zone:	10
Latitude:	48 54 02	Northing:	5416850
Longitude:	124 10 22	Easting:	414050
Elevation:	1000 m		
Comments:	Turn N off Highway #18 at Cowichan Lake dump 2 km W of Lake Cowichan village. Follow Meade Creek ML 12 km until the road climbs steeply up Mt Franklin. Main showing is in SW corner of Osirus A claim.		
Commodities:	Rhodonite, gemstones, manganese, copper		

MINERALS

Significant:	Rhodonite, jasper, rhodochrosite, spessartine, pyrite
Associated:	Chalcedony, magnesite, calcite, quartz
Comments:	Manganese oxides.

The Rocky deposit is located on the S slopes of Mount Franklin, about 3.5 km N of Youbou. The Rocky workings have not been located, but are believed to be in the vicinity of the Osirus A claim.

Rhodonite and jasper occur in lenticular masses in cherts and cherty tuffs, with associated rhodocrosite and spessartine. Disseminated pyrite and chalcopyrite occur in quartz veins associated with diorite. Rhodonite found in areas of dark ribbon chert which may be cut by major faults.

The main showing on the Osirus A claim is deep pink rhodonite which compares favourably with Hill 60 (092B 027). Rhodonite occurs locally in bands 2 mm to 5 mm wide and in crackle breccia veinlets and lenses.

Surface stripping was done on the Rocky claim in 1977 – 1978; 555 kg of rhodonite were produced. Gem quality rhodonite on Osirus A is low.

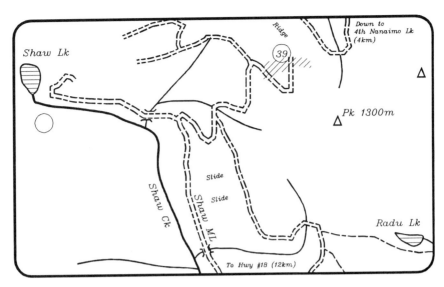

39. MINFILE NUMBER: 092F 563
NAME(S): FLIGHT 5, FLIGHT

* * **Gemstones**

Status:	Showing		
NTS Map:	092F 01W 092C 16W	UTM Zone:	10
Latitude:	49 01 19	Northing:	5430650
Longitude:	124 24 43	Easting:	396775
Elevation:	820 m		
Comments:	Approximate centre of the jasper body.		
Commodities:	Silica, manganese, magnetite, copper		

MINERALS

Significant:	Jasper, pyrite, magnetite, pyrrhotite, chalcopyrite
Associated:	Quartz
Comments:	Massive pyrrhotite was reported but not located on maps.

Although the Flight 5 showing is located in the Upper Nanaimo River valley, not far from the now defunct Green Mountain Ski Resort, the best current access is from Cowichan Lake and the East Shaw ML. Caution: this was an active logging road at the time of this writing. Consult the logging company, stationed at Mesachie Lake. From Youbou, drive 14 km W, then 12 km N on E Shaw Creek. Some kilometres before Shaw Lake, as the road curves W, cut up the switchbacks. Jasper deposit spans the NW ridge of the peak for 1300 m, and down the W side.

An extensive jasper body containing minor magnetite occurs at the McLaughlin Ridge Formation/Nitinat Formation contact. It is reported to be 10 m to 15 m thick, traceable for 250 m, dipping vertically, and is hosted in basaltic rocks overlain by sandstones and siltstones. The jasper is locally broken with minor infillings of magnetite and is laterally succeeded by lenses, blocks or wedges of jasper with minor pyrite. These are overlain by fine-grained chloritic tuff, laminated cherty tuff and finally by hematitic altered lapilli tuff. A 30 cm wide associated shear zone contains chlorite, kaolinite, sericite, pyrite, trace chalcopyrite and malachite.

There is a report of a showing on west Shaw Creek (092F 186) of a rhodonite and quartz deposit (Lat:49 00 05, Long:124 25 45 / N: 5428375, E: 395468, elevation 575 m). Lenses of manganese silicates, mainly rhodonite, occur in highly folded red and white cherty tuffs. The lenses are occasionally coated with hard, black siliceous oxide.

The mineralization is exposed over an area measuring 100 m by 30 m. Tests in 1940 showed samples contained pyrolusite, rhodochrosite and rhodonite, which are the oxide, carbonate and silicate of manganese, respectively.

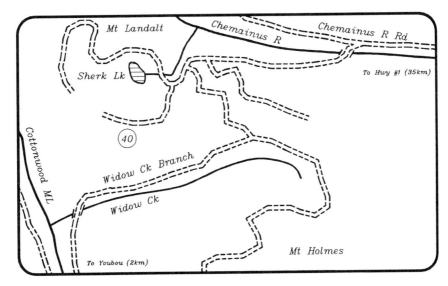

40. MINFILE NUMBER: 092C 026   * Gemstones
NAME(S): SHERK LAKE, SHERK, STRIKER, MALJO CREEK

Status:	Showing		
NTS Map:	092C16E	UTM Zone:	10
Latitude:	48 55 17	Northing:	5419200
Longitude:	124 12 36	Easting:	411350
Elevation:	1060 m		
Comments:	Located 800 m S of Sherk Lake at top (W) end of Chemainus River.		
Commodities:	Rhodonite, gemstones, manganese		

MINERALS

Significant:	Rhodonite, rhodochrosite, jasper, pyrite, pyrrhotite

The Sherk Lake showing is located about 800 m S of Sherk Lake, which can be reached from Youbou by taking the Mainline Road on E side of Widow Creek. After 1 km, road turns E into upper Widow Creek and at 5 km turns left up to Sherk Lake. Alternate approach is up Chemainus River's Copper Canyon road (long but flat). About 35 km up the valley, leave the main gravel road and cross from N to S side of river, and a km later turn left up to Sherk Lake. Note: at time of press, the Chemainus River road was closed at Rheinhart Lake turnoff.

Rhodonite, rhodochrosite and jasper occur in cherts and cherty tuffs. Rhodonite development is restricted to dark ribbon chert and it may be cut by major faults. Pyrite and pyrrhotite also occur in the area hosted by felsic tuffs.

A cherty tuff bed contains a jasper horizon which hosts irregular, lenticular masses of rhodonite and rhodochrosite. The jasper horizon has been traced along strike for more than 300 m and is up to 1 m wide. The largest lens is several centimetres wide and 30 cm to 60 cm long.

CHEMAINUS VALLEY

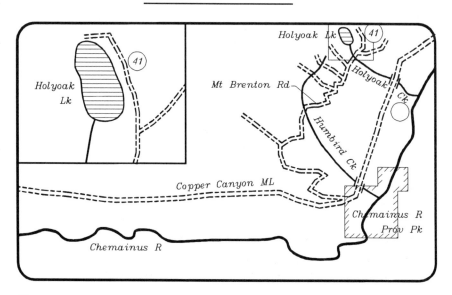

41. MINFILE NUMBER: 092B X06   *** Gemstones
NAME(S): HOLYOAK LAKE

Status:	Showing		
NTS Map:	092B13	UTM Zone:	10
Latitude:	48 54 15	Northing:	5416750
Longitude:	123 50 10	Easting:	43900
Elevation:	1060 m		
Comments:	Old borrow site, located on NE side of Holyoak Lake, close to road.		
Commodities:	Feldspar porphyry		

MINERALS

Significant:	Flowerstone

The Holyoak Lake showing is located just NE of the lake near the wall. Access is from the Copper Canyon road up Chemainus River. From Highway #1, drive 11 km. Just after Humbird Creek sign turn right (N) and take right forks all the way. Showing is around a road borrow site on NE side of lake. The mid-section of this road requires 4WD.

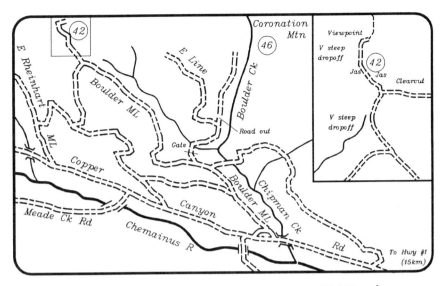

42. MINFILE NUMBER: 092C X08
NAME(S): DEASON 2

** Minerals

Status:	Showing		
NTS Map:	092C 16E	UTM Zone:	10
Latitude:	48 57 00	Northing:	5422250
Longitude:	124 03 19	Easting:	422750
Elevation:	0950 M		
Comments:	Located at the NW end of the Boulder ML. A red jasper dyke crosses the road. Elsewhere, jasper lenses in green host rock make spectacular contrasts.		
Commodities:	Iron, magnetite		

MINERALS

Significant:	Magnetite
Associated:	Jasper, chalcedony

The Deason 2 showing is located near the end of the Boulder ML about 0.5 km before it turns a corner and ends at a viewpoint down to Rheinhart Lake. The jasper horizon is exposed over a 2 m width and along strike for approximately 50 m. The material dips 80 degrees N and strikes 220 degrees. The jasper is bright red, with quartz stringers and blebs.

Some 50 m S, the road cut has exposed jasper lenses up to 50 cm in length interbedded in banded green cherty material.

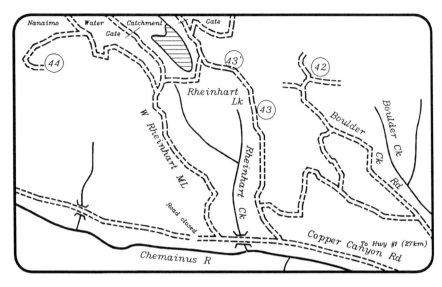

43. MINFILE NUMBER: 092C 144
NAME(S): SOGNIDORO

*** Gemstones

Status:	Prospect		
NTS Map:	092C16E	UTM Zone:	10
Latitude:	48 57 28	Northing:	5423100
Longitude:	124 04 37	Easting:	421950
Elevation:	700 m		

Comments: Located 2 km S of Rheinhart Lake on E access road from Chemainus River. Jasper horizons contain finely disseminated and massive magnetite lenses.

Commodities: Gold, silver, copper, magnetite, gemstones

MINERALS

Significant: Chalcopyrite, pyrite, bornite, galena, sphalerite, chalcanthite, magnetite, molybdenite

Associated: Quartz, carbonate

Alteration: Azurite, malchite, hematite

The Sognidoro showing is located some 2 km S of Rheinhart Lake on the E side of the valley, and extends upslope for some distance. Significant bright red jasper is exposed at the road. A 100 m adit appears on the nearby Trek claims, possibly from as early as 1918. Pyrite, chalcopyrite, hematite, and magnetite appear within the jasper horizons. Galena was observed in a quartz vein cutting a diabasic outcrop within the southerly flowing creek on the W side of the claim.

The main vein is the McDougall vein, striking 320 degrees and dipping 70 degrees E, which has been traced for 265 m. Mineralization apparently increases in quantity toward the N end of the vein. A sample, from a pit on the vein, contained iron oxides, malachite and chalcanthite.

Two jasper showings are located in the central claim area. The horizons are exposed over 30 m and 25 m widths and along strike for 200 m and 50 m respectively. The jasper appears to occur in lenses but it could be part of a continuous horizon displaced by right-lateral faulting. The jasper, brick to scarlet red with metallic gray patches, is cut by numerous quartz veinlets (up to 5 mm). Iron oxides and malachite stain locally. Pyrite and chalcopyrite occur primarily in the veinlets. Finely disseminated and massive magnetite occurs within the jasper lenses.

About 1 km S of Rheinhart Lake on the same road, quartz has intruded steeply dipping shale beds alongside the road. Well defined cubic pyrite crystals are visible within the host matrix.

Thanks to Ms Winnie Espitalier for first reporting this site.

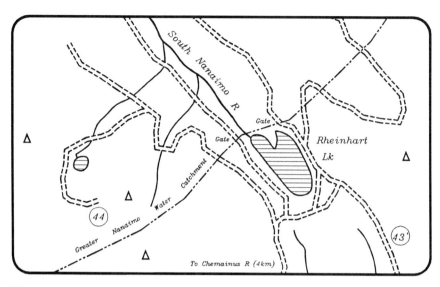

44. MINFILE NUMBER: 092C X01   *** Gemstones,
NAME(S): OLD NICK   Geology

Status:	Showing		
NTS Map:	092C 16E	UTM Zone:	10
Latitude:	48 57 34	Northing:	5423200
Longitude:	124 06 45	Easting:	418500
Elevation:	1050 m		
Comments:	Located on logging road S above mountain lake.		
Commodities:	Iron, magnetite		

MINERALS

Significant:	Magnetite
Associated:	Jasper, chalcedony
Comments:	Jasper horizons interbedded in cherty sediments and felsic flows (very rare in Fourth Lake Formation). Significant jasper boulders.

Thanks to Dr Nick Massey for first reporting this site.

LADYSMITH AREA

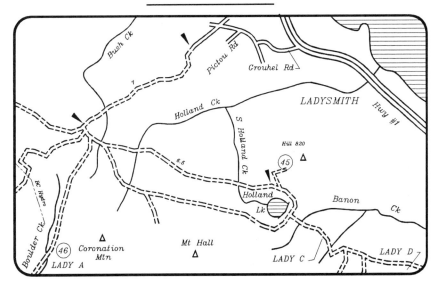

45. MINFILE NUMBER: 092C X08 ** Geology
NAME(S): HOLLAND LAKE QUARRY

Status:	Past producer		
NTS Map:	092C13W	UTM Zone:	10
Latitude:	48 57 20	Northing:	5422750
Longitude:	123 51 45	Easting:	437100
Elevation:	740 m		
Comments:	Short track climbs from the N end of Holland Lake dam wall toward summit of Hill 820 m.		
Commodities:	Iron		

MINERALS
Significant:	Pyroxene lavas, magnetite
Associated:	Breccia, jasper, pyrite, chalcopyrite

This site, once used as a source of quarry material for the construction of the Holland Lake dam, offers a rare chance to see pyroxene-rich pillow lavas and breccias of the Nitinat Formation (Paleozoic era) alongside gabbors and granodiorites of the lower to middle Jurassic Island Intrusions (known as the Ladysmith Pluton).

Thanks to Dr Nick Massey for first reporting this site.

If you follow the road SE from the bottom of the dam wall, it crosses Banon Creek almost immediately, then 3 km later cuts W along Banon Creek before climbing S upslope for 1 km. Take the left fork and traverse across slope 3 km to LADY D (092B 076, Lat: 4854 43, Long: 123 47 54) on the NE flank of Mt Brenton, overlooking the town of Chemainus. An old magnetite dump and adit show massive magnetite breccia containing up to 20% pyrite. Quartz and jasper are present in the footwall, allong with chalcopyrite.

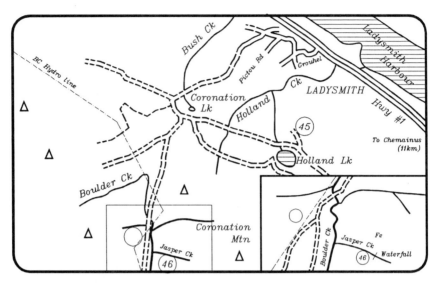

46. MINFILE NUMBER: 092B 029
NAME(S): LADY A

** Gemstones,**
Minerals

Status:	Developed prospect		
NTS Map:	092B13W	UTM Zone:	10
Latitude:	48 55 27	Northing:	5419250
Longitude:	123 57 05	Easting:	430300
Elevation:	540 m		

Comments: Located to the W of the upper Chipman (Boulder) Creek, 3 km due W of Coronation Mountain. The deposit spans the valley, with showings on both sides of the creek. Access from Chemainus River valley and up Chipman Creek, but the road is de-activated about 2km from Jasper Creek. Alternative is off Highway #1 just N of Ladysmith onto Grouhel Rd., then onto back road to Holland Lake before turning S onto Boulder Creek ML. BC Hydro lines keep the bottom of the valley open.

Commodities: Iron, magnetite

MINERALS

Significant: Magnetite

Associated: Jasper, chalcedony

The Lady A is a stratabound taconite deposit composed of gray chert and red jasper hosting bands of very fine-grained magnetite with minor specularite. The deposit consists of two lenses which outcrop along strike for 100 m, with widths up to 18 m. An average thickness, determined from drilling, is less than 9 m. Estimated reserves are 360,000 tonnes grading 25% iron.

Considerable jasper float may be found in Chipman (Boulder) Creek. This may be traced to a dyke at stream level, 400 m up Jasper Creek (200 m below the waterfall). To the N, an open slope exposes magnetite.

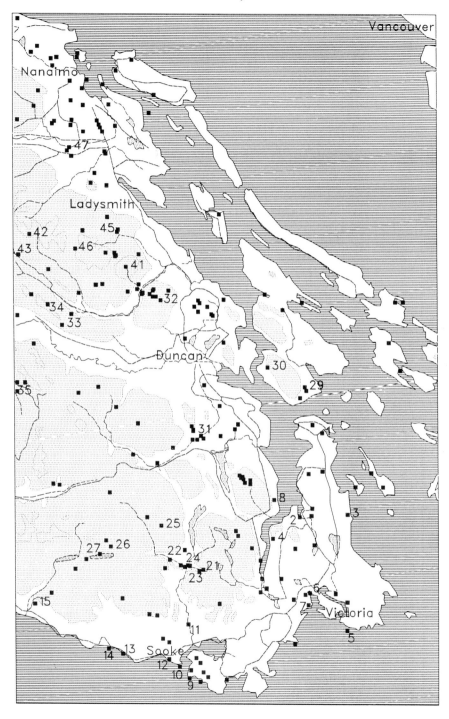

MAP 1

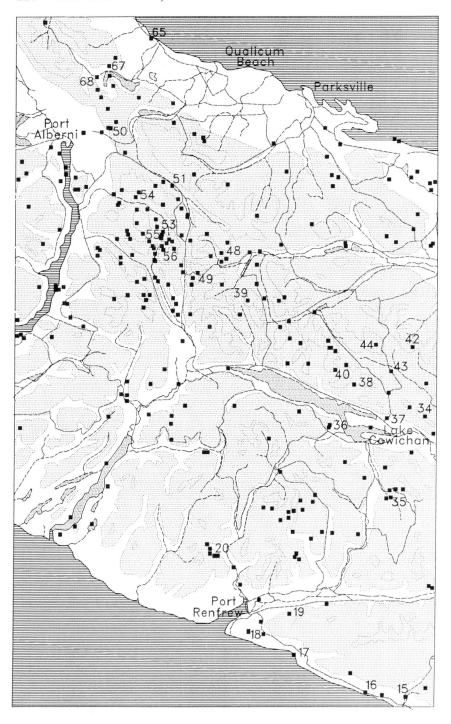

MAP 2

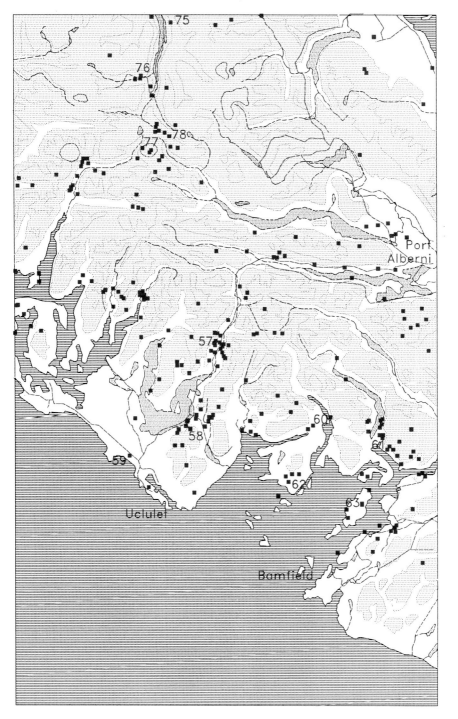

MAP 3

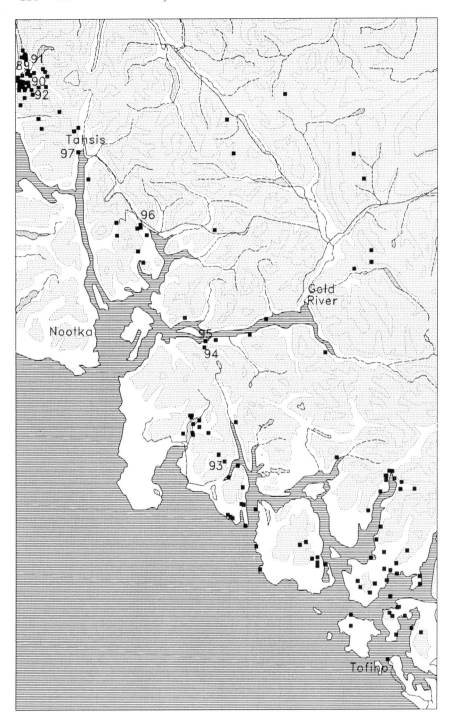

MAP 4

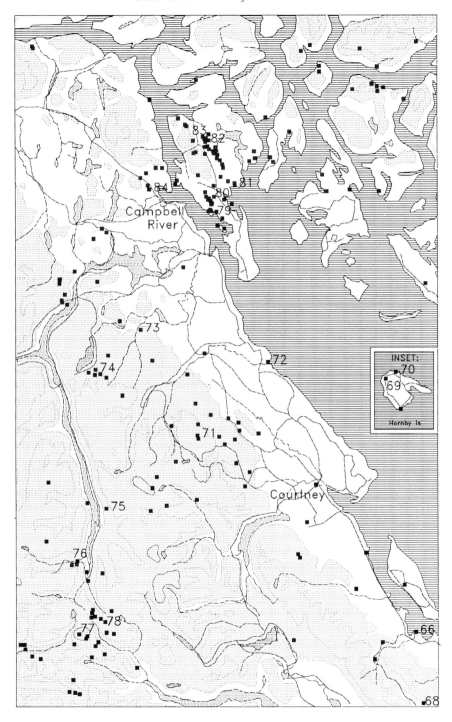

MAP 5

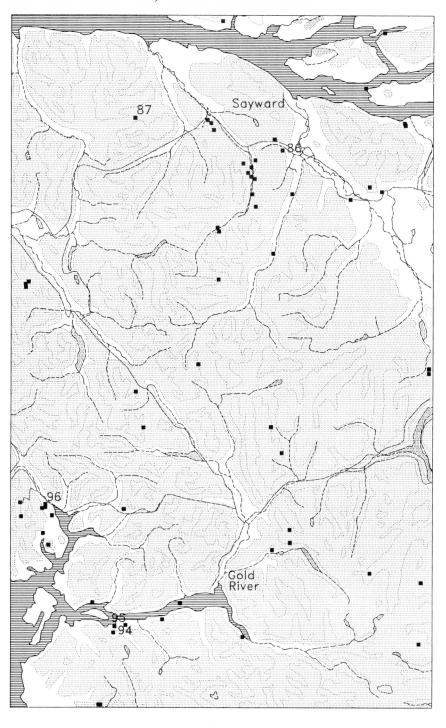

MAP 6

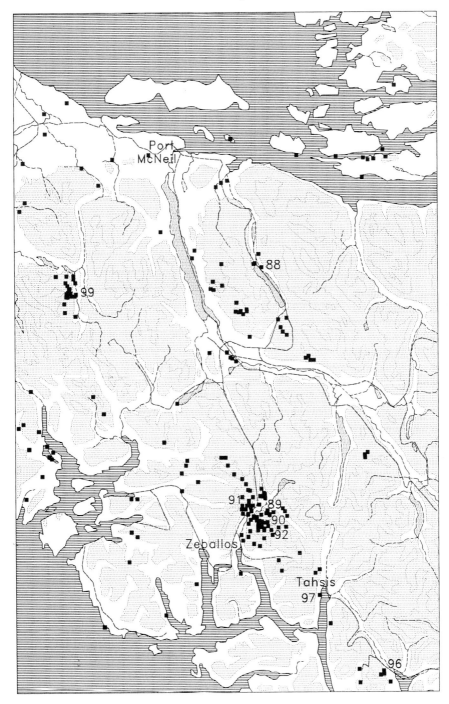

MAP 7

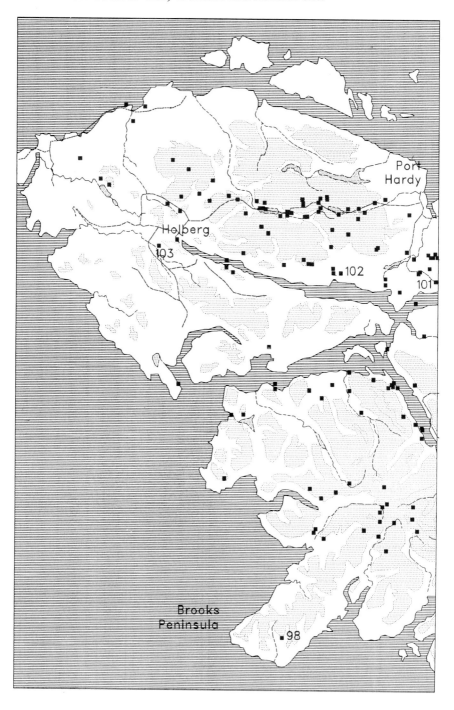

MAP 8

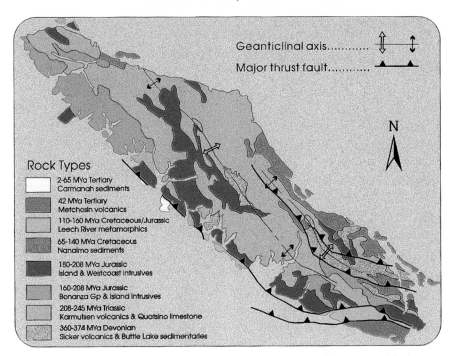

Simplified geological map of Vancouver Island, showing principal surface rocks and faults. (Courtesy of the BC Geological Survey Branch)

Minerals are found both on the earth's surface and below. Spectacular limestone caves at Horne Lake and further north offer rare glimpses beneath the surface. (Photo courtesy of Mr. P. Curtis)

Cubic pyrite crystals in volcanic tuff, past producer Island Copper Mine, Port Hardy. See Site 101.

Crystalline pyrite on druzy quartz, Mt Washington. See Site 71.

Chalcopyrite sample from past producer Mt Washington Copper Mine. See Site 71.

Blue-green malachite and azurite staining on copper sulfides from past producer Sunro Mine, Jordan River. See Site 15.

Druzy quartz crystals form in fissures near Port Alberni. See Site 52.

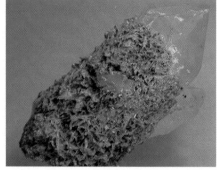

A clear quartz crystal (8 cm long) overgrown with green epidote crystals, found near Sayward. See Site 87.

Druzy quartz crystal is found in the Port Alberni area, close to fault lines.

Pegmatite from the Walker Creek area shows mica and black tourmaline crystals. See Site 27.

Iron-stained red jasper, hematite and quartz found on southern Salt Spring Island. See Site 30.

Banded red chert from the upper Chemainus River, near Sherk Lake. See Site 40.

Dense barite banded with celestite from the past producer Brynnor Mine. See Site 58.

Acicular crystals of gray-black stibnite (antomony sulfide) in gangue.

Old style metal pan showing 'colour' (fine gold powder). See Site 17.

Plastic gold pan, loupe and pick-tools with gold nuggets. See Site 37.

Massive hematite stratum exposed at Villalta, Upper Nanaimo River, Site 48. (Photo courtesy of Dr. N. Massey)

Flowerstone porphery shows dark gabbro host with white feldspar crystals at Holyoak Lake, Site 43. (Photo courtesy of Dr. N. Massey)

Gemstone rhodonite from the famous Hill 60 Ridge in the Cowichan Valley, cut into cabochons or tumble-polished. See Sites 29, 33, 38. (Photo courtesy the Ed Travers Collection)

Columnar rhyolite outcrop shows evidence of rapid cooling after volcanic extrusion, near Mt Ozard.

Blue gemstones are rare. Imperial-blue dumortierite from the past producer Island Copper Mine, Port Hardy. See Site 101.

River-worn red jasper pebble on green chert bed, Kammat Creek. See Site 51.

Volcanic breccia exposed in roadcut close to Villalta, Upper Nanaimo River. See Site 48.

Heavily folded schists near Walker Creek show evidence of powerful metamorphic processes. See Site 27.

Fossil of branching coral *retrophyllia* in Sutton Limestone, Lake Cowichan. See Site 36. (Photo courtesy of Dr. N. Massey)

Marine fossil shells found near Muir Creek, west of Sooke. See Site 14. (Photo courtesy of Dr. R. Hebda)

Site of the famous Hollings rhodonite quarry (now closed) on Salt Spring Island. See Site 29.

The past producer Sunro Mine lies deep within the Jordan River gorge. See Site 15.

Massive jasper 'float' found downhill from hematite-quartz dyke, Salt Spring Island. See Site 30.

View looking north over Horne Lake to Mt Mark's limestone cliffs. (Photo courtesy of Dr. N. Massey)

Upper Bamberton limestone quarry, close to the Malahat Pass. See Site 8.

Along the old Port Alberni railway line, a cutting reveals heavily folded and banded material. See Site 50.

NANAIMO RIVER VALLEY

The Upper Cretaceous coal-bearing sediments of the Nanaimo Group along the E coast of Vancouver Island overlay Permian and Triassic volcanics. Five different basins have been identified. The Nanaimo Basin runs from Crofton to Nanoose Bay; the Comox Basin stretches from Nanoose Bay to Campbell River. These are the only two that have proven to be economically viable to date. Coal is exposed on the Iron River (see site 73) at two locations. On Chute Creek, at the S end of the Comox Basin, three seams of about 30 cm each have been seen. N of the Puntledge River, both Dove Creek and Brown River have exposed seams up to1.5 m thick. A seam outcrops for about 6 km (although partly beneath overburden) between Coal Creek at the E end of Comox Lake, and the Puntledge River. It was mined to a limited extent near Coal Creek and in the vicinity of old Chinatown. See also Port Hardy, sites 101-102.

The city of Nanaimo gave its name to the Nanaimo Group of sediments, formed in the late Cretaceous period, when the earlier groups were weathering away into a shallow, tropical sea. The formation is the source of extensive coal and fossil deposits.

Coal was first reported in the area in 1849. Production in the Nanaimo coalfield was from three major seams: the Wellington, the Newcastle and the Douglas. The total workable area was 19.3 km long and averaged 1.5 km in width.

From 1836 to 1968, about 75 million tonnes of coal were produced on Vancouver Island. Most went to California. When petroleum was discovered in 1902, the market changed radically, and the operators had to look elsewhere for customers. In 1911, 4,676 men worked in the mines, including 714 Chinese and Japanese. Miners earned $3.30 to $5 per day. The worst of many accidents was the Nanaimo Colliery explosion of 1887 which killed 148 men.

There is a current belief that all the coal on Vancouver island is worked out. Not so. The latest estimates suggest there are more confirmed unmined reserves in the strip between Nanaimo and Campbell River than has been mined in the last 150 years! Unfortunately, much of what remains is either shallow and dirty (full of clay and shales) or deep and clean (expensive to reach). The only operational mine at present is the Quinsam Mine (see Site 73) W of Campbell River.

WARNING: Do not enter old coal mines! The weak host rock makes them very prone to "settling" unexpectedly. A sudden settlement could ruin your whole day. Second, and more importantly, methane gas builds up in shafts. Undetectable, it has taken the lives of countless miners and more than a few amateurs who ventured into unknown territory. Stay clear! Stick to fossicking on mine dumps. It's messy, but safe.

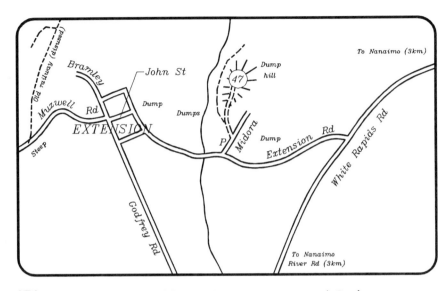

47A MINFILE NUMBER: 092GSW028   * Geology
NAME(S): EXTENSION

Status:	Past producer, underground		
NTS Map:	092G 04W	UTM Zone:	10
Latitude:	49 05 57	Northing:	5438702
Longitude:	123 57 30	Easting:	430042
Elevation:	200 m		
Comments:	It is difficult today to find surface seams in the Nanaimo area. Of the many underground operations that brought material to the surface, the ones around Extension are typical. Abandoned workings, slag heaps, old mining equipment and history. The Nanaimo Mining Museum near Cedar (SE of Nanaimo) is worth a visit.		
Commodities:	Coal		

MINERALS

Significant:	Coal

Coal seams ranged from 0.6 m to 0.9 m in thickness and were overlain by a bed of mudstone ranging from 25 cm to 76 cm in thickness. In places as much as 25 cm of excellent quality coal lies above the mudstone. The mine was abandoned in January, 1958 due to persistent shaly coal.

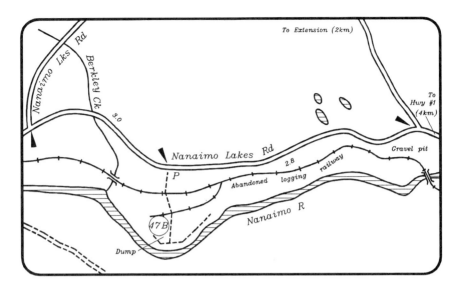

47B MINFILE NUMBER: 092G 043   *** Geology**
NAME(S): WHITE RAPIDS

Status:	Past producer, underground		
NTS Map:	092G 04W	UTM Zone:	10
Latitude:	49 04 06	Northing:	5435278
Longitude:	123 57 41	Easting:	429776
Elevation:	100 m		
Comments:	Located on the attractive Nanaimo River, the area has great walks as well as being full of mining history.		
Commodities:	Coal		

MINERALS

Significant:	Coal

The White Rapids mine was developed as part of the Wellington seam which occurs in the Upper Cretaceous Extension Formation of the Nanaimo Group. The seam was usually between 60 cm and 75 cm of highly volatile bituminous rank coal. The seam was characterized by an extremely soft shale roof.

The mine operated between 1943 and 1950, and closed on account of the thinness of the seam and other factors. It produced 256 thousand tonnes of coal.

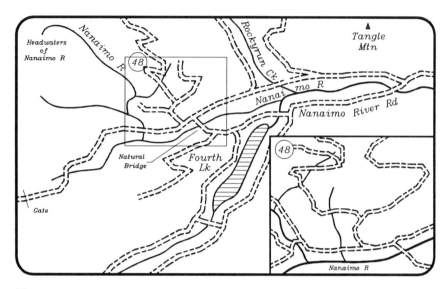

48. MINFILE NUMBER: 092F 384   *** * Minerals**
NAME(S): VILLALTA

Status:	Developed, prospect		
NTS Map:	092F 01W	UTM Zone:	10
Latitude:	49 05 25	Northing:	5438316
Longitude:	124 28 20	Easting:	392510
Elevation:	850 m		
Comments:	Red hematite hardpan developed on sulphide-rich skarn.		
Commodities:	Gold, silver, zinc, copper, lead		

MINERALS

Significant:	Pyrite, chalcopyrite, hematite, phyrrhotite, arsenopyrite, marcasite, galena, magnetite
Associated:	Siderite, calcite, quartz, ilvaite
Alteration:	Serpentine, goethite

The Villalta occurrence area is underlain by volcanics, clastic sediments and limestone. Poorly sorted conglomerates and hematitic mudstones overlie the limestones.

The rocks include volcanic breccia, tuff, andesite, argillite and chert. The overlying crinoidal limestone, with minor chert and tuff, likely belongs to the Mount Mark Formation, and are tightly folded.

Extensive areas of powdery hematite with gold values occur at the top of the limestone and measures 110 m by 30 m by 14 m. Randomly distributed massive sulphide bodies, comprised of pyrite, pyrrhotite, chalcopyrite, arsenopyrite, marcasite and minor galena and magnetite, which occur in the limestone, are the likely source for the hematite zone.

Thanks to Dr Nick Massey for first suggesting this site.

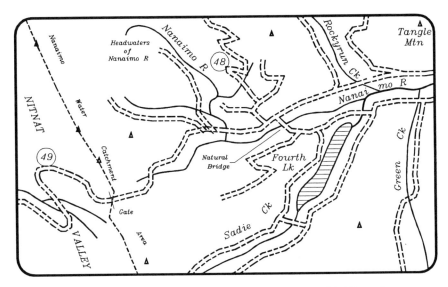

49. MINFILE NUMBER: 092F X04
NAME(S): MASSEY

** Minerals

Status:	Showing		
NTS Map:	092F 02W	UTM Zone:	10
Latitude:	49 41 00	Northing:	5435200
Longitude:	124 01 12	Easting:	388600
Elevation:	800 m		
Comments:	Well developed augite crystals (up to 1 cm) in pillow lavas.		

MINERALS

Associated: Augite

Access is by way of the Upper Nanaimo River basin. A gate closes the road at the watershed at the W end of the valley, where the road drops down to the Nitinat River. At the first hairpin bend, where the road switchbacks S, augite rich pillow basaltic lavas are found in the road cutting and upslope. Site is about 1 km W of the gate at the watershed. Well-formed crystals visible.

Thanks to Dr Nick Massey for first reporting this site.

PORT ALBERNI AREA

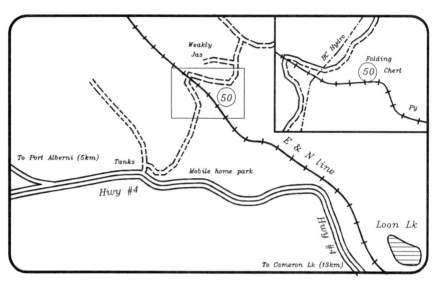

50. MINFILE NUMBER: 092F 451
NAME(S): MAIN, RAILWAY

** Minerals,
Geology

Status:	Showing		
NTS Map:	092F 07W	UTM Zone:	10
Latitude:	49 16 35	Northing:	5458500
Longitude:	124 43 55	Easting:	374000
Elevation:	320 m		

Comments: Easy access along a disused railway line. Two cuttings exhibit different features: impressive folding (synclines and anticlines) at the first (W) site; sulfides at the second (E).

Commodities: Agate

MINERALS

Significant: Pyrite, chert

Park where the road crosses the railway line. The first site is about 300 m E of this. Well-defined folding of the host matrix characterizes the western cutting. A thin but clearly defined seam of gray quartz (chert) 5 cm wide follows the fold.

At the E railway cutting (200 m further), coarse-grained massive pyrite occurs in seams and pods over an area 10 m by 7 m on a vertical rock-cut face. The pods are contorted and irregular in shape and up to 10 cm by 50 cm by 100 cm in size. The host rock consists of fine to medium-grained, diabase-gabbro which contain magnetite and pyrrhotite. Veinlets are common throughout the rock, but are most concentrated near the pyrite pods. Malachite is associated with some veinlets.

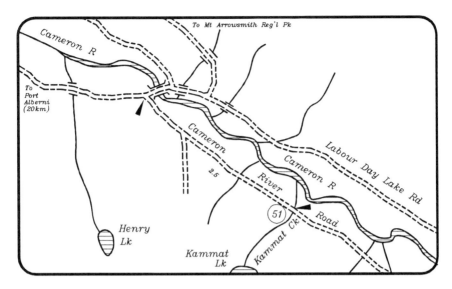

51. MINFILE NUMBER: 092F 233
NAME(S): KAMMAT CREEK

★ ★ Gemstones

Status:	Showing		
NTS Map:	092F 02E	UTM Zone:	10
Latitude:	49 11 14	Northing:	5449250
Longitude:	124 35 01	Easting:	384600
Elevation:	520 m		
Comments:	From the Coombes–Port Alberni highway, turn S into Cameron River about 5 km past Cathedral Grove Provincial Park, following the signs to the Arrowsmith ski area. When ski road crosses left over river, continue straight (S). 10 km from the end of the blacktop on a good gravel road along the Cameron River brings you to Kammat Creek on right (W) side.		
Commodities:	Copper, gold, silver		

MINERALS

Significant:	Pyrite, jasper, magnetite, copper
Associated:	Druzy quartz, siderite

The Kammat Creek showing is located 19 km SE of Port Alberni. Pyritic jasper with magnetite and minor black chert also hosts mineralization in the area.

The Cameron River Fault runs at least 60 km from Horne Lake in the N to the Upper Nanaimo River basin. Many of the side creeks in the Cameron River valley cut this fault and the resultant contact zones. The Cop Creek showing (similar) is located 1 km W. Several other small creeks adjacent to Kammat Creek showed similar quartz crystals and druzes.

Large boulders of brecciated jasper are reported in the valley, extending as far SE as Labour Day Lake. The river shows jasper float.

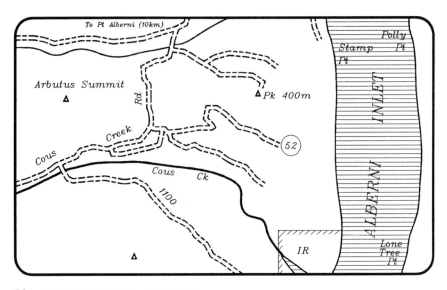

52. MINFILE NUMBER: 092F X05
NAME(S): ARBUTUS

Status:	Showing		
NTS Map:	092F 02W	UTM Zone:	10
Latitude:	49 12 20	Northing:	5420755
Longitude:	124 50 18	Easting:	348600
Elevation:	330 m		
Comments:	Great views overlooking the Alberni Inlet.		
Commodities:	Quartz		

MINERALS

Significant:	Quartz

Druzy quartz reported found on 400 m high (unnamed) hill W of Stamp Narrows in Alberni Inlet. Take Cous Creek Road SE from Sproat Lake. From the Cous Creek bridge, go back 2 km (E) to Arbutus Line E. When road forks, take upper (L) road to extreme end. Site is on the road.

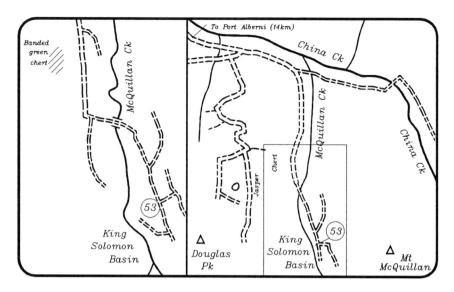

53. MINFILE NUMBER: 092F 429   ** Gemstones
NAME(S): MCQUILLAN CREEK, MCQUILLAN

Status:	Showing		
NTS Map:	092F 02E	UTM Zone:	10
Latitude:	49 07 41	Northing:	5442725
Longitude:	124 37 05	Easting:	381950
Elevation:	920 m		

Comments: Road deteriorating annually. Impressive valley with giant headwalls surround the site. Although jasper float was found during recent field searches, no source was discovered by the author. The lode is reported alongside the road on the upper terrace in King Solomon Basin. A new road (1996) takes off from where Duck Main leaves China Creek Main, and heads SSE into upper King Solomon Basin.

Commodities: Iron, gemstones

MINERALS

Significant: Jasper

Associated: Quartz

The McQuillan Creek showing is located 17 km SE of Port Alberni up China Creek, on the creek of the same name. The rocks comprise hematitic jasper, basalt flows, hematitic basalt breccia, feldspar porphyry, basalt intrusives, basalt and tuff.

Hematitic jasper is exposed in a 1.7 m wide by 15 m long outcrop, close to the road, trending approximately 155 degrees. The jasper consists of 75 to 90% bright brick-red jasper with 10 to 20% interstitial clear quartz containing about 5 to 10% very fine-grained disseminated hematite. Irregular hematite-filled fractures, up to 2 cm thick, crosscut the jasper. Locally the jasper contains massive hematite bands, 0.5 to 1.5 m thick.

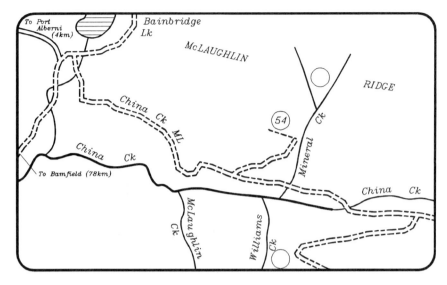

54. MINFILE NUMBER: 092F 331   ** Minerals
NAME(S): 900

Status:	Developed prospect		
NTS Map:	092F 02E	UTM Zone:	10
Latitude:	49 10 15	Northing:	5447564
Longitude:	124 40 00	Easting:	378511
Elevation:	650 m		
Comments:	On the N side of China Creek, 15 km from Port Alberni, up Mineral Creek.		
Commodities:	Gold		

MINERALS

Significant:	Gold, pyrite, arsenopyrite
Associated:	Quartz, jasper, carbonate, magnetite

Beneath and cross-cutting the chert horizon is a quartz vein stockwork. Native gold, pyrite, magnetite and arsenopyrite occur in quartz veinlets in the chert and jasper, as well as in narrow carbonate veinlets.

Drilling in 1988 intersected quartz stockworks with visible gold. Several old trenches and a shaft explore the mineralized zone, which is found close to the tree line at the top of the clearcut. Jasper float and druzy quartz has spilled down onto the track (4WD, and washed out at time of press).

Some 200 m higher up Mineral Creek, past producer Debbie (MINFILE: 092F 079) shows similar minerals, and across China Creek to the S, developed prospect Regina (MINFILE: 092F 078) on the E side of Williams Creek at 600 m altitude continues the structure.

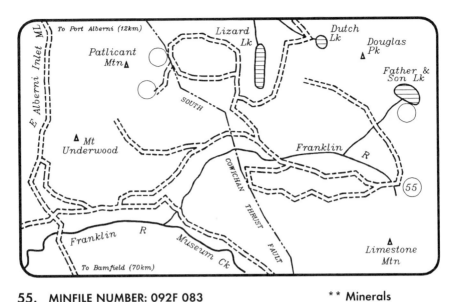

55. MINFILE NUMBER: 092F 083   **Minerals
NAME(S): THISTLE

Status:	Past producer, underground		
Map:	092F 02E	UTM Zone:	10
Latitude:	49 06 24	Northing:	5440382
Longitude:	124 38 08	Easting:	380625
Elevation:	800 m		
Comments:	Adits and gloryholes.		
Commodities:	Gold, silver, copper		

MINERALS

Significant:	Pyrite, chalcopyrite, magnetite, pyrrhotite
Associated:	Quartz, calcite, diopside, epidote
Alteration:	Diopside, epidote, malachite

The Thistle mine is located about 16 km SE of Port Alberni, just S of Father and Son Lake. Basaltic flows and pillow basalt of the Karmutsen Formation are underlain by a succession of volcanics and sediments of the Sicker and Buttle Lake groups. These include basaltic flows, agglomerates and bedded tuffs, and limestones and marbles of the Mount Mark Formation.

Disseminated to massive sulphide mineralization, consisting of pyrite, chalcopyrite and minor pyrrhotite plus sulphide rich quartz-carbonate veins, occur in sheared pyritic quartz-sericite schists with mafic volcanic flows and tuffs. A nearby limestone, which strikes 170 degrees and dips 65 degrees SW has largely been replaced by diopside (skarn). Disseminated magnetite, some of which has been oxidized to hematite, occurs in the calcite. Malachite occurs in places.

Two ore zones, 40 m apart, measure 2 m to 20 m long by 1 m to 8 m wide. The Thistle Mine was reported by early workers to be a skarn deposit in altered limestone, intruded by fine-grained diorite.

Of a geological interest, the South Cowichan Thrust Fault is well exposed to the W of Lizard Lake. Driving back from Thistle to Port Alberni down the Franklin River, after 4 km a road turns N up to Lizard Lake. The Fault is visible at the extreme W end of the loop road, W of the lake. A road heading SW from there across the SE flank of Patlicant Mountain exposes a hornblende feldspar dacite sill. Thanks to Dr Nick Massey for first describing this site.

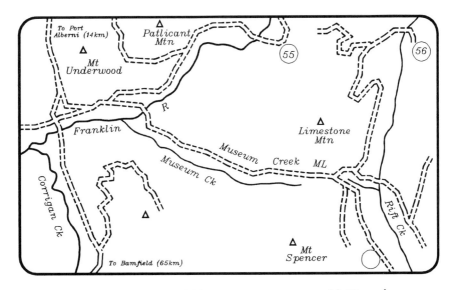

56. MINFILE NUMBER: 092F 084
NAME(S): BLACK PANTHER

* * Minerals

Status:	Past producer, underground		
NTS Map:	092F 02E	UTM Zone:	10
Latitude:	49 06 00	Northing:	5439596
Longitude:	124 36 25	Easting:	382697
Elevation:	900 m		

Comments: Located up the Franklin River and Museum Creek. After leaving the Port Alberni-Bamfield Rd at Franklin River, drive 9.5 km E. At the pass, turn left (N) up Rift Creek. Another 4 km brings you out on the S slopes of Mt McQuillan to the old mine site.

Commodities: Gold, silver, lead, copper, zinc

MINERALS

Significant: Pyrite, chalcopyrite, galena, sphalerite
Associated: Quartz
Alteration: Ankerite, quartz

The Black Panther mine, discovered in 1936, is about 20 km SE of Port Alberni. In the area, the N Cowichan Fault strikes NNE, separating andesites from diorite. Quartz veins, lenses, stockworks and stringers containing variable amounts of sulphides, mainly pyrite, chalcopyrite, minor galena and sphalerite occur in a shear zone which sub-parallels the andesite/diorite contact. The wallrock is strongly altered by ankeritic carbonate for widths of several centimetres up to 9 m. The main shear zone, which has been traced for at least 3 km, is locally cut by quartz stringers that are up to 1 m wide and 12 m long.

Another site of interest is in the Mt. Mark Formation limestone E of Mt Spencer on the W Rift Creek ML (Mt Spencer – MINFILE: 092F 409).

There, crinoidal limestone is apparent. Crinoids (popularly known as sea lilies or feather stars) flourished in warm seas during the Paleozoic. Looking much like a thin-stemmed wine glass, the base attached to the sea bed, while a flexible stem supported up to 20 short arms that formed the cup or calyx. The arms filtered food from the passing water.

UCLUELET-TOFINO AREA

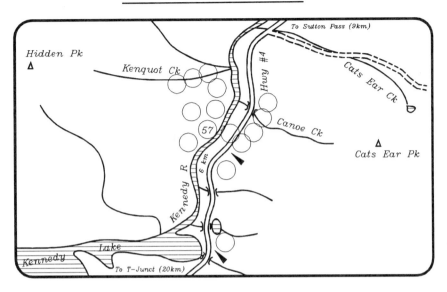

57. MINFILE NUMBER: 092F 029/033/044/045/046/051/052 etc
NAME(S): CAPTAIN HOOK, GIANT BEAR   * Minerals

Status:	Showings		
NTS Map:	092F 03W	UTM Zone:	10
Latitude:	49 09 25	Northing:	5447800
Longitude:	125 25 20	Easting:	32370
Elevation:	400 m		
Comments:	Some sixteen showings or prospects have been identified close to Highway #4, all within a few square kilometers of each other. Centred about a point 15 km W of Sutton Pass (the watershed W of Sproat Lake), on the Ucluelet/Tofino road.		
Commodities:	Copper		

MINERALS

Significant:	Chalcopyrite, bornite
Associated:	Quartz, calcite

Copper mineralization occurs at many localities over a distance of about 2 km, near Highway #4 and Kennedy River. Mineralization within the volcanics includes chalcopyrite, specularite, bornite, malachite and azurite.

Where Kennedy River flows into Kennedy Lake an area has been set aside for goldpanning and rockhounding on the gravel bars.

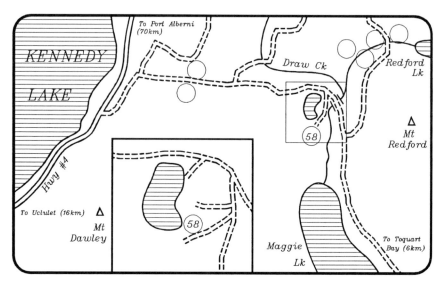

58. MINFILE NUMBER: 092F 001   *** Minerals
NAME(S): BRYNNOR

Status:	Past producer		
NTS Map:	092F 3	UTM Zone:	10
Latitude:	49 02 35	Northing:	5425275
Longitude:	125 26 00	Easting:	335850
Elevation:	50 m		
Comments:	Worked-out iron deposits.		
Commodities:	Iron, agate		

MINERALS

Significant: Hematite, quartzitic greenstone (cave marble), barite, celestite

Shortly after Highway #4 drops down to Kennedy Lake, take the Toquart Bay Road (E). The site (now a flooded pond) is some 3 km from the Highway #4 turnoff. Just beyond, several tracks lead right to the tailings pile. A green cave is found, along with abundant hematite, and dense barite/celestite gangue.

Numerous other mineral sites exist in the area. Most are located along a 2 km spur road that runs N off the Toquart Bay Road, 1 km W from the Brynnor site. The Maggie Mine (092F 005) yielded fine epidote specimens. Epidote crystals are also found along Highway #4 rock cuttings where it climbs away from Kennedy Lake, heading towards Port Alberni.

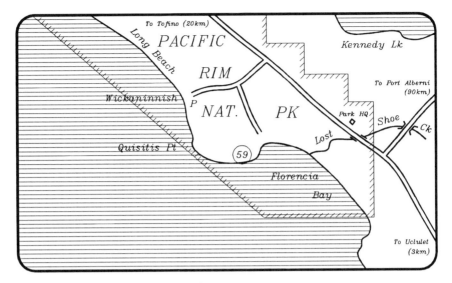

59. MINFILE NUMBER: 092C 121   ** Gemstones
NAME(S): WRECK BAY, BLACK SAND, FLORENCIA BAY

Status:	Past producer, open pit		
NTS Map:	092C13E	UTM Zone:	10
Latitude:	48 59 40	Northing:	5429950
Longitude:	125 37 29	Easting:	308000
Elevation:	5 m		
Comments:	Located at the base of the cliffs on Florencia Bay.		
Commodities:	Gold		

MINERALS

Significant:	Gold, magnetite, silica

The Wreck Bay (now Florencia Bay) beach placers occur between Kennedy Lake and the W coast of Vancouver Island. The placers extend from Ucluelet to Tofino Inlet on a flat coastal plain composed of unconsolidated sands, fine gravels and thin beds of blue clay.

These Pleistocene sediments contain a small amount of black sand (magnetite) and fine gold which is being continually concentrated at the base of the cliffs along the bay. Prospectors and campers have historically panned the sand periodically. The amount of sand is small and the quantity of gold is very small where concentration due to wave action has not taken place. The gold probably comes from the quartz veins that occur to the W of Kennedy Lake.

Located inside the Pacific Rim National Park, 5 km from the T junction at the end of Highway #4. Turn down to the sea (Wickaninnish) and after 1 km turn left to Florencia Bay.

BARKLEY SOUND

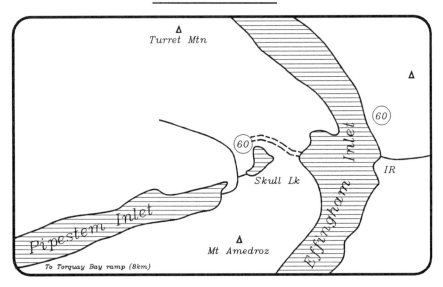

60. MINFILE NUMBER: 092F 414   * Minerals
NAME(S): PIPESTEM INLET, EFFINGHAM INLET

Status:	Past producer, open pit		
NTS Map:	092F 03E	UTM ZONE:	10
LATITUDE:	49 02 38	NORTHING:	5434400
LONGITUDE:	125 10 57	EASTING:	340500
Elevation:	50 m		
Comments:	Centred on limestone band between Pipestem and Effingham Inlets, 1 km in from Effingham Inlet, on Skull Lake. Boat approach.		
Commodities:	Limestone, marble		

MINERALS

Significant:	Calcite

A band of limestone extends E-NE for 5 km from the E end of Pipestem Inlet to the W shore of Effingham Inlet. The band widens from 500 m at Pipestem Inlet to 2.0 km at Effingham Inlet.

At Pipestem Inlet, the band comprises fine-grained, blue limestone, while at Effingham Inlet, it consists of white crystalline limestone cut by numerous dykes. In thin section the limestone at Effingham Inlet displays bands of reddish-brown ferruginous mud and unidentified fossil structures cemented by recrystallized calcite and fine limy mud. It was quarried for marble until 1902.

On the E side of Effingham Inlet is a high reddish-brown bluff, having a compact, fine-grained texture. The rock consists of angular grains of quartz, which are cemented together by a fine aggregate of granular material, which is almost wholly hematite. The rock is reported to be impure jasper (possible building stone). There are intrusions of a greenish volcanic rock having an amygdaloidal texture. Listed as MINFILE 092F 428.

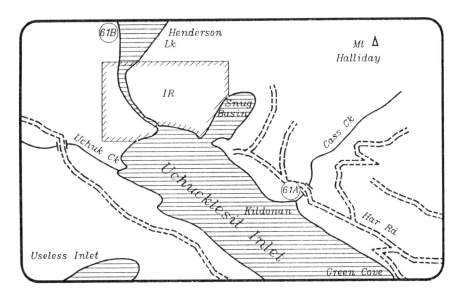

61. MINFILE NUMBER: 092F 157
NAME(S): CASCADE

* Minerals

Status:	Past producer, open pit		
NTS Map:	092F 03E	UTM Zone:	10
Latitude:	49 00 27	Northing:	5430000
Longitude:	125 00 24	Easting:	353250
Elevation:	20 m		
Comments:	Small tonnage mine at Kildonan on Uchucklesit Inlet.		
Commodities:	Copper, gold, silver		

MINERALS

Significant:	Chalcopyrite, pyrrhotite
Alteration:	Garnet, epidote, hornblende, quartz

The Cascade occurrence is located within a few hundred metres of the beach at Kildonan, Uchucklesit Inlet. A diabase dyke (andesite) intrudes limestone and is impregnated with chalcopyrite and iron pyrite. The deposit is also reported to be associated with skarn material made up of garnetite, epidote, hornblende and quartz and is described as a vein.

The deposit was mined in 1904 – 1905 and produced 113 tonnes of ore. Most of it was taken from an open cut on the surface showing.

NW of this site, near the summit bluff on the W side of Henderson Lake, there is a chalcopyrite showing (61B on the map). Listed as MINFILE 092F 165, BIG BLUFF, Lat: 49 02 06, Long: 125 03 03, UTM Northing 5433150, Easting 350100, elev 100 m. A vein of copper ore is reported in an adit 6 m long and 60 cm wide, striking 25 degrees and dipping 26 degrees to the SE. The vein of copper (described as a ledge of ore) can be traced on surface for at least 12 m.

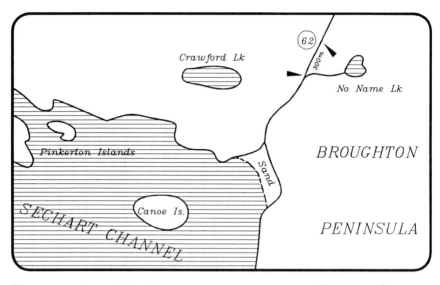

62. MINFILE NUMBER: 092C 065   * * * Minerals
   NAME(S): SECHART

Status:	Prospect		
NTS Map:	092C 14E	UTM Zone:	10
Latitude:	48 57 38	Northing:	5425275
Longitude:	125 14 33	Easting:	335850
Elevation:	50 m		
Comments:	Source of cinnabar material.		
Commodities:	Mercury		

MINERALS
Significant: Cinnabar, mercury

The Sechart mercury deposit is located on the N shore of Sechart Channel. Exact location uncertain. The deposit, known since 1890, had undergone considerable development by 1911. More development work was conducted from 1916 to 1917 and again in 1928 and 1969. To date, the property is developed by about 100 m of drifts and two shafts, one 10 m in depth and the other 5 m. The geology of the area is dominated by coarse-grained quartz diorite, containing slices of metamorphic rock. In the immediate vicinity of the workings the rocks consist of andesitic greenstone, gabbro and minor limestone. The deposit occurs in ankeritized and silicified greenstone. Mineralization consists of patches and streaks of red cinnabar and native mercury. The mercury mineralization penetrated along fault zones and fractures with associated silica flooding and sericite, chlorite and ankerite alteration.

Further N upslope at about the 300 m level, several magnetite prospects and developed prospects (092C 002/003/004/005) occur extensively, mixed with fine-grained rock and pyrite.

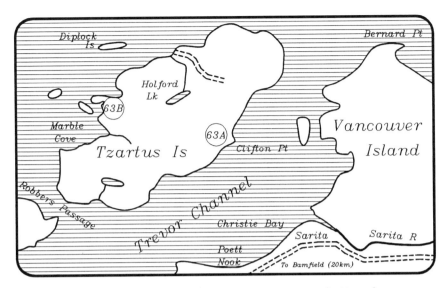

63. MINFILE NUMBER: 092C 033
NAME(S): MOUNTAIN

* **Minerals**

Status:	Developed prospect		
NTS Map:	092C14E	UTM Zone:	10
Latitude:	48 55 23	Northing:	5420755
Longitude:	125 04 00	Easting:	348600
Elevation:	250 m		
Comments:	Mineralized outcrop at highest point, 800 m W of Clifton Point.		
Commodities:	Iron, magnetite		

MINERALS

Significant:	Magnetite, chalcopyrite, pyrite
Associated:	Garnet

The Copper Island grants cover an area about 1.5 km W from and including Clifton Point on the W coast of Tzartus Island, Barkley Sound. Most of mineralization occurs on the Mountain claim, 800 m from Clifton Point.

Work recorded from 1899 to 1901 comprised trenching, a 4 m shaft, and a 30 m tunnel. Approximately 1800 tonnes was mined from the main showing. In 1961, diamond drilling was completed. The ore bodies are hosted in interbedded andesitic tuffs and tuffaceous sediments.

The main showing (largely mined out) is an irregular body of massive magnetite, roughly 12 m by 20 m, close to the contact. It grades into irregular lenses of magnetite (up to 10 cm thick) interlayered with garnet or quartz-epidote skarn. Garnet and minor chalcopyrite and pyrite impregnate the lenses. Other showings are exposed at higher elevations to the E.

The Marble Cove Limestone (MINFILE: 092C 132) outcrops on the W side of Tzartus Island. Up to 750 m wide, it extends NE from Marble Cove for 1000 m.

LASQUETI ISLAND

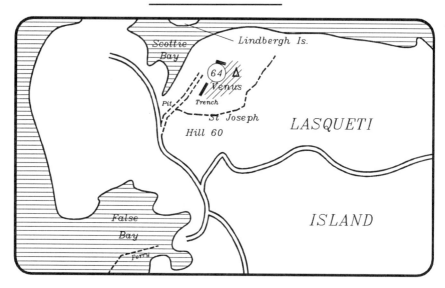

64. **MINFILE NUMBER: 092F 131, 133**   * **Minerals**
     NAME(S): VENUS, JUNEAU

Status:	Past producer, underground		
NTS Map:	092F 09W	UTM Zone:	10
Latitude:	49 30 41	Northing:	5484940
Longitude:	124 19 56	Easting:	403560
Elevation:	10 m		
Comments:	Portal No. 2 on the N end of Lasqueti Island at the head of Barnes Cove between Lindbergh and Jelina islands, 2.5 km N-NE from the village of Lasqueti (False Bay).		
Commodities:	Copper, gold, silver, zinc		

MINERALS

Significant:	Chalcopyrite, pyrite, magnetite, sphalerite
Alteration:	Chlorite

Lasqueti Island is dominated by dark, gray-green amygdaloidal basalt. Narrow shear zones along the stock margins contain minor quartz veining.

The principal mineralization occurs as a shoot of massive chalcopyrite with minor pyrite, up to 1.2 m wide, with associated magnetite and sphalerite. Past work consists of several adits, some underground development and extensive surface workings which are approximately 350 m W of the adits.

Juneau (092F 133) is a prospect just E of Scottie Bay Road, 750 m S of Scottie Bay and 1.5 km N of Lasqueti (False Bay). Approximately 600 m along a N-NE strike, several trenches explored three narrow, parallel quartz veins. Shear zones in quartz diorite near the contact with basalt contain disseminated chalcopyrite, chalcocite, bornite and pyrite.

PARKSVILLE-CAMPBELL RIVER

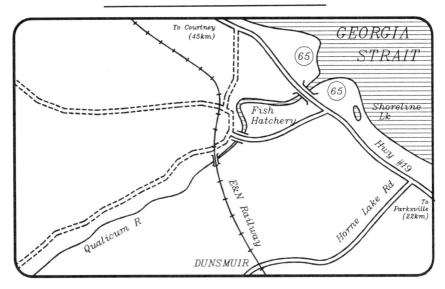

65. MINFILE NUMBER: 092F X01
NAME(S): QUALICUM BAY, DUNSMUIR BEACH

** ** Gemstones**

Status:	Showing		
NTS Map:	092F 07E	UTM Zone:	10
Latitude:	49 24 00	Northing:	5472800
Longitude:	124 37 00	Easting:	38300
Elevation:	2 m		
Comments:	Mouth of the Qualicum River		
Commodities:	Dallasite, jasper		

A popular holiday site, the island's east coastline offers more than just swimming, camping and great views across the Strait of Georgia to the Coast Mountains. Both the river and the adjacent beaches are reported to show dallasite and red jasper, washed down from the Horne Lake showings (092F X02). Easy to get to and a fun way to spend an afternoon with small children, turning over barnacle-encrusted pebbles looking for the tell-tale blush of deep-red jasper, or the green and white patterns of dallasite. To clean your specimens, scrape most of the seaweed and seashells off them and leave them in dilute bleach for a day to kill that "fishy" smell. Let them dry in the sun and scrub clean with a wire brush.

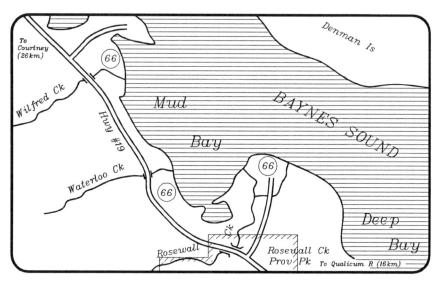

66. MINFILE NUMBER: 092F X03   ** Gemstones
NAME(S): ROSEWALL CREEK, MUD BAY

Status:	Showing		
NTS Map:	092F 07E	UTM Zone:	10
Latitude:	49 27 00	Northing:	5480000
Longitude:	124 43 40	Easting:	37050
Elevation:	2 m		
Comments:	Mouth of Rosewall Creek.		
Commodities:	Dallasite, flowerstone porphyry		

Both the river and the adjacent beaches are reported to show abundant dallasite. There are quite wide sea grass and mud flats. A good place to take children, but wear gumboots, and watch the tide as it can come in surprisingly quickly. An old pail or ice cream bucket makes a good collecting tool and limits the amount of "good stuff" you have to drag back to the car.

Much of the flowerstone is washed up from a deposit on Texada Island across the Strait of Georgia. It is not easy to identify when encrusted with marine life - small 'flowers' of feldspar set in a dark basalt matrix often look like barnacles from a distance.

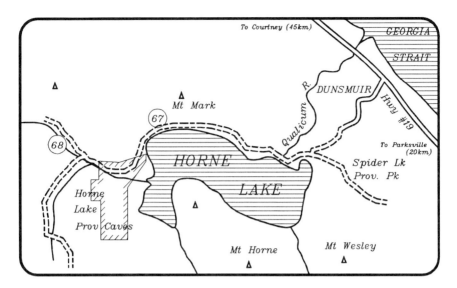

67. MINFILE NUMBER: 092F 089
NAME(S): HORNE LAKE

* Minerals, geology

Status:	Showing		
NTS Map:	092F 07E	UTM Zone:	10
Latitude:	49 21 35	Northing:	5468660
Longitude:	124 43 47	Easting:	374400
Elevation:	400 m		
Comments:	Centre of outcrop on S side of Mount Mark.		
Commodities:	Limestone		

MINERALS

Significant:	Calcite

The deposit is a limestone bed with exposed thicknesses up to 360 m, as revealed on the S face of Mount Mark, N of Horne Lake. The deposit consists of medium to light gray, fine to coarse-grained recrystallized, yet well bedded bioclastic limestone, containing abundant crinoid remains. Thin sections display numerous whole and fragmented crinoid discs in a very fine grained limy mud matrix with minor secondary silica. At Mount Mark, the limestone contains minor thin chert beds in the upper and lower portions of the exposed section. In the middle of this unit, the limestone is interbedded with lenses and beds of argillite and tuff. Several gabbro sills intrude the limestone near the top of the section.

From Island Highway, follow signs for Horne Lake Caves Provincial Park.

68. MINFILE NUMBER: 092F X02   ** Gemstones
NAME(S): UPPER QUALICUM RIVER

Status:	Showing		
NTS Map:	092F 07W 092F 07E	UTM Zone:	10
Latitude:	49 21 16	Northing:	5468200
Longitude:	124 46 20	Easting:	371300
Elevation:	400 m		
Comments:	Dallasite breccia placers. Some 2 km beyond the Caves Provincial Park car park, the road turns left across a wooden bridge. Instead, continue straight along the true lefthand bank of the river for 1 km.		
Commodities:	Dallasite		

MINERALS

Significant:	Unknown
Associated:	Quartz

Dallasite (a quartz-epidote-basalt breccia) is reported found in the gravel bars about a kilometre upstream of where the Qualicum River flows into Horne Lake. Large pieces with well-defined white quartz filling are easily seen in the shallow river. One advantage of this locale over site 65 and site 66 is the absence of barnacles that mask the true colours. Look for pieces that have sharp colour edges. Real collector's pieces are those having re-crystallized rectangular basalt patterns. These formed in the white material after the liquid quartz penetrated and melted the green-brown basalt, and then cooled.

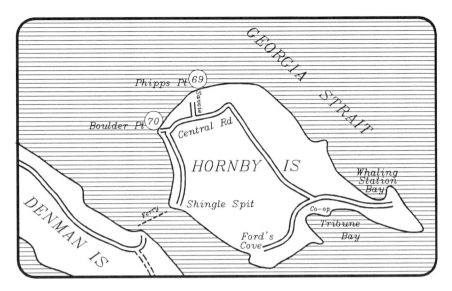

69. MINFILE NUMBER: 092F X06
NAME(S): PHIPPS POINT

** Fossils

Status:	Showing		
NTS Map:	092F 10E	UTM Zone:	10
Latitude:	49 42 15	Northing:	490000
Longitude:	124 40 00	Easting:	37700
Elevation:	2 m		
Comments:	About 2.5 km from ferry terminal. Turn down steep road and park next to ancient concrete wharf.		
Commodities:	Fossils		

Ammonite and baculite fossils are found in concretions in the clay cliffs about a kilometer E of Phipps Point (and the old concrete dock), as well as on the rock beach below the cliffs when the tide is out. Take a hammer, newspaper and collector's bag, and wear gumboots. Only visit sites 69 and 70 during spring low tides, when the whole point is relatively dry. Watch for returning tides; they rise quickly.

When the tide is out, a series of gently SE dipping shale and sandstone beds are exposed. As you walk from Boulder Point to Phipps Point, you are travelling forwards in time, as the layers become younger and younger. Search for gray nodules, ranging in size from tennis balls to footballs. Pry these out and crack them open. Often they are empty, but occasionally you will find one worth collecting. Wrap in newspaper and store carefully: fossils are fragile.

70. MINFILE NUMBER: 092F X05   ** Fossils
NAME(S): BOULDER POINT

Status:	Showing		
NTS Map:	092F 10E	UTM Zone:	10
Latitude:	49 43 00	Northing:	5488500
Longitude:	124 39 35	Easting:	37600
Elevation:	2 m		
Comments:	Turn off Central Rd onto Savoie Rd. Park at end. Footpath leads 500 m to the beach. Best at very low tide, when the entire point is exposed. Gumboots handy. Once you are on the beach, look back and make a note of where the path goes back up the bank. It is not obvious and many people miss it completely when they return.		
Commodities:	Fossils		

Ammonite and baculite fossils are found in concretions in the clay cliffs at Boulder Point, as well as on the rock beach below when the tide is out. A heavy hammer is recommended to split the concretions to determine whether they contain fossils. When you are successful, make a note of the strata, and work along the line, as it is likely there will be others in the same era. Watch for incoming tides; there are several places where the sea reaches the foot of the cliffs, and you may have to wade (or worse).

What to do with old fossils when you are bored with them? Please don't throw them out! Contact either the Vancouver Island Paleontological Museum Society in Qualicum Beach, or the Courtney Museum in Courtney. They will be delighted to receive your contribution, and your name will be associated with the specimens forever.

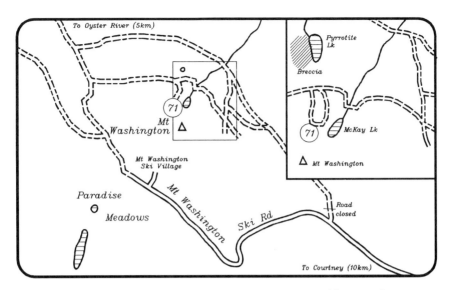

71. MINFILE NUMBER: 092F 117 　　　　　　　**** Minerals**
NAME(S): MOUNT WASHINGTON COPPER

Status:	Past producer, open pit		
NTS Map:	092F14W	UTM Zone:	10
Latitude:	49 45 49	Northing:	5514669
Longitude:	125 18 03	Easting:	334300
Elevation:	1295 m		

Comments: Open pit, 0.8 km upslope (W) of McKay Lake and 1.4 km N of Mt Washington. Because of its altitude, access in late summer (Jul-Oct) months only.

Commodities: Copper, gold, silver, arsenic, molybdenum, zinc, lead

MINERALS

Significant: Chalcopyrite, pyrite, bornite, covellite, realgar, orpiment, molybdenite, sphalerite, galena, pyrrhotite, arsenopyrite

Comments: Gold and silver mineralogy not known.

Associated: Quartz

The Mount Washington Copper deposit is located on a ridge on the N side of Mount Washington. The mineralization is contained in a 1.5 m to 7.5 m wide tabular zone that contains a stockwork of chalcopyrite-pyrite-quartz veins, and disseminated chalcopyrite in the sediments and the sill. Low gold and silver values are associated with the veins. Bornite, covellite, realgar, orpiment, pyrrhotite, arsenopyrite, molybdenite, sphalerite and galena are present. Between 1964 and 1967, 381,773 tonnes of ore were mined from two open pits, producing 131 kg of gold, 7,235 kg of silver and 3,548 tonnes of copper.

Just to the S, developed prospect DOMINEER (092F 116) at 1,400 m altitude, exhibits extensive mineralization of a similar nature. Steep climb SW from McKay Lake, but great views.

The main work areas have recently been capped and contoured, but interesting material can still be found. Gilles Lebrun reports there are records of bornite, calcite, chalcedony, chrysocola, chalcocite, duranusite, goethite, hematite, hessite, lapidocrocite, limonite, malachite, magnetite, marcasite, pararealgar, siderite, stibnite, tetrahedrite and wherlite. Another report suggests extremely rare cooperite (PtS) crystals have been found.

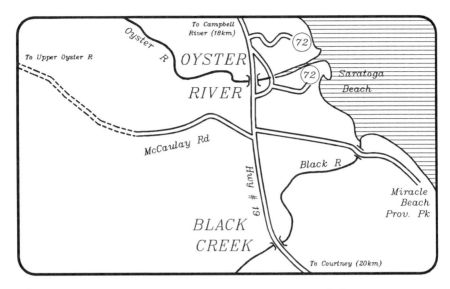

72. MINFILE NUMBER: 092F X07   *** Gemstones**
       NAME(S): OYSTER RIVER

Status:	Showing		
NTS Map:	092F14W	UTM Zone:	10
Latitude:	49 52 00	Northing:	5526400
Longitude:	125 07 00	Easting:	34800
Elevation:	20 m		
Comments:	Located on Oyster River, on east side of Island Highway #19 bridge.		
Commodities:	Flowerstone porphyry, dallasite		

MINERALS

Significant:	Gabbro
Associated:	Feldspar

The river shows flowerstone porphyry boulders and pebbles along a kilometer of its length from the sea, and along the adjacent beaches. A good place to take children to beachcomb. The river mouth is deep, so you cannot get from the N side to the S, and vice versa. Choose a side, and spend an hour or so with a collecting bucket. Flowerstone has small, pale, flower-like patterns (feldspar) set in a dark green or black material (basalt) and is often difficult to see because most rocks with barnacles look like flowerstone.

QUINSAM LAKES

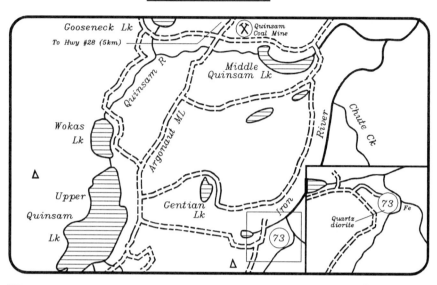

73. MINFILE NUMBER: 092F 076   * Minerals
    NAME(S): IRON RIVER

Status:	Developed prospect		
NTS Map:	092F13W	UTM Zone:	10
Latitude:	49 56 08	Northing:	5526229
Longitude:	125 29 10	Easting:	317131
Elevation:	457 m		
Commodites:	Iron, limestone, magnetite		

MINERALS
Significant: Magnetite
Associated: Pyrite, epidote, calcite

A significant limestone bed 1.2 km wide extends from the SE shore of Upper Quinsam Lake for about 6 km to the headwaters of the Iron River, dipping NE. Magnetite and chalcopyrite are concentrated at the N end of a skarn, adjacent to a quartz diorite at Iron River. Two exposures are reported. The W one is 80 m by 30 m, formed on the W bank in a horsehoe bend of the river. The E deposit is at river level, on the E bank of the Iron River, and consists mainly of high purity magnetite.

POINT OF INTEREST

Coal occurs on Vancouver Island in the late Cretaceous Nanaimo Series. Major deposits extend from Mud Bay S of Courtney to Campbell River, along the plains bordering Georgia Strait, and inland up to 20 km.

Quinsam Coal (096F 319, Lat: 49 56 08, Long: 125 29 10), just N of Middle Quinsam Lake is one of only two operating mines on Vancouver Island

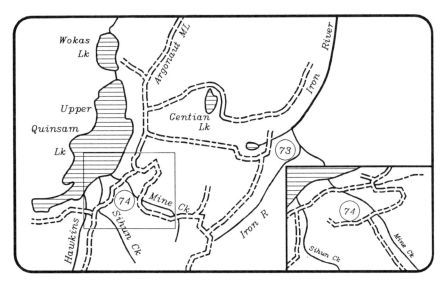

74. MINFILE NUMBER: 092F 075
NAME(S): IRON HILL

* Minerals,
geology

Status:	Past producer, open pit		
NTS Map:	092F13E	UTM Zone:	10
Latitude:	49 51 45	Northing:	5526229
Longitude:	125 32 40	Easting:	317131
Elevation:	457 m		
Comments:	Open pit. The old iron mine is a potential producer of garnet, presently under review.		
Commodities:	Iron, limestone, garnet, magnetite		

MINERALS

Significant:	Magnetite
Associated:	Garnet, pyrite, epidote, calcite

Turn off the Campbell River — Gold River Highway #28 2 km W of Echo Lake (20 km from Campbell River) onto the Argonaut Mainline. Mine is 16 km from turn-off, on S side of Upper Quinsam Lake, on Mine Creek.

The Argonaut mine is a massive magnetite-magnetite/garnetite skarn situated on Iron Hill, just E of upper Quinsam Lake. Skarn mineralization occurs along the contact between limestone and pillowed basalts and consists of massive garnetite and magnetite with minor amounts of epidote, calcite, and pyrite.

Between 1951 and 1957, over 3.5 million tonnes of ore were mined, from which 2 million tonnes of concentrate was shipped. The tailings and waste pile contain fine-grained magnetite and garnet.

About 1 km W of Iron Hill, massive magnetites in greenish volcanics show on Sihun Creek, with associated chalcopyrite and pyrite, interlaced by quartz and calcite veins. There are reports of cobalt showings close to Upper Quinsam Lake, between Sihun and Mine Creeks.

STRATHCONA PROVINCIAL PARK

The centre of Vancouver Island features Strathcona Provincial Park. Cutting the many layers of the island's terrane is Buttle Lake. Like a giant knife, it allows the visitor to look down and back in history from the recent ice ages of 18,000 years ago at the top, right back to the Devonian era (408-362 million years ago) when the West Coast complex (quartz diorite, gabbro and gneiss) were laid down. These layers are visible at the S end of the lake, at shoreline. It is in this layer that we find the Myra Falls/Lynx mine, one of the two active mines on Vancouver Island.

For more information on this area, the Geological Survey Branch of the BC Ministry of Energy, Mines & Petroleum Resources produces an excellent large-format geological map of the region, titled "Geology of Strathcona Provincial Park." See Chapter 10 for address.

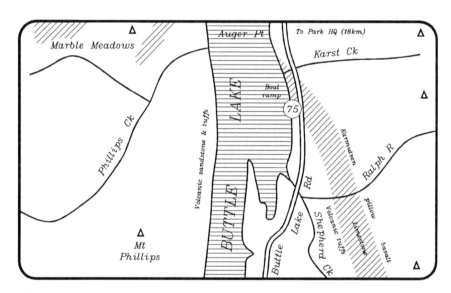

75. MINFILE NUMBER: 092F 422   * Geology
      NAME(S): BUTTLE LAKE

Status:	Past producer, open pit		
NTS Map:	092F12E	UTM Zone:	10
Latitude:	49 39 13	Northing:	5502950
Longitude:	125 31 15	Easting:	318050
Elevation:	225 m		
Comments:	Location centred on quarry on E side of Buttle Lake, about 21 km S of Park Warden's offices.		
Commodities:	Limestone		

MINERALS
Significant: Calcite

A band of limestone extends for 11 km along the E side of Buttle Lake, near its S end. It is unconformably overlain to the E by basaltic flows. Underlying volcanic breccia, tuff and argillite outcrop to the W. The band is underlain near its S end by a gabbro sill. The unit dips gently NE to SE. Several faults segment the limestone.

On a point of interest, it was research carried out in the black basalts on the E side of the lake which established the age and origin of what is today Vancouver Island (Wrangellia). Scientists at the Geological Survey of Canada took rock cores that, when analyzed, showed the formation was laid down about 230 million years ago, somewhere near today's Hawaii. The plate carrying the new underwater range moved slowly NE, and likely hit the North American continent about 100 million years ago.

Further N at Lupin Falls (8km S of Park HQ) the dark basalts of the Karmutsen Formation are visible as pillow lavas low down across the lake.

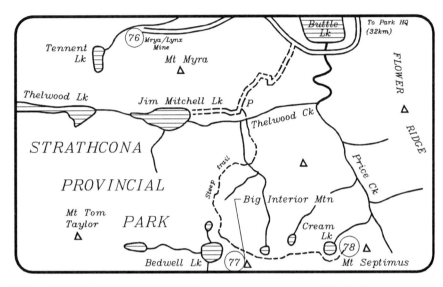

76. MINFILE NUMBER: 092F 071   ** Geology
NAME(S): MYRA FALLS, LYNX MINE

Status:	Producer, open pit, underground		
NTS Map:	092F12E	UTM Zone:	10
Latitude:	49 34 04	Northing:	5493613
Longitude:	125 36 13	Easting:	311743
Elevation:	427 m		
Comments:	The Lynx portal is located 0.5 km N of Myra Creek, 3 km W of Buttle Lake. Follow lake road to its end. If in doubt, stop and listen: the mine's extractor fans can be heard for miles.		
Commodities:	Copper, zinc, lead, gold, silver, cadmium		

MINERALS

Significant:	Chalcopyrite, sphalerite, galena, pyrite, tennantite, bornite, stromeyerite, digenite, covellite
Associated:	Quartz, sericite, chlorite, talc, pyrrhotite, barite

The major ore zones of the Lynx mine are located within an area of 2.5 km by 0.7 km. The massive sulphide horizon lies within a zone of quartz-feldspar rhyolite tuff and minor chert, and comprise chalcopyrite, galena, sphalerite, pyrite and pockets of barite. Minor tennantite, bornite, pyrrhotite, digenite, covellite and stromeyerite are present. The lenses are up to 12 m thick and 244 m long, pinching out along strike.

The Lynx occurrence was mined by open pit methods from 1966 to 1976, then by underground mining techniques to the present. Significant new finds recently have ensured an extended life for the mine. For mine tours, contact the Chief Geologist, Westmin Resources Ltd., P.O.Box 8000, Campbell River, BC V9W 5E2, or fax: 250-287-7123.

POINT OF INTEREST

The only other operating mine on Vancouver Island is presently Quinsam Coal in the Quinsam Lake district NE of Myra Falls. (See Site 73: IRON RIVER for details.)

---

77. MINFILE NUMBER: 092F 090
    NAME(S): BIG INTERIOR MOUNTAIN
        ** Geology, fossils**

Status:	Showing		
NTS Map:	092F 05E	UTM Zone:	10
Latitude:	49 27 22	Northing:	5481130
Longitude:	125 34 13	Easting:	313740
Elevation:	1680 m		
Comments:	Centre of limestone outcrop. Rugged foot access.		
Commodities:	Limestone		

MINERALS

Significant:	Calcite
Associated:	Silica
Comments:	As chert within the limestone.

The deposit consists of a mass of limestone. The outcrop covers a 1.25 km by 2.5 km area on Big Interior Mountain, S of Bedwell Lake. The limestone is bounded to the W and N by granodiorite and quartz diorite and partially overlain to the S by pillowed basalts. The deposit is comprised of light gray to white recrystallized limestone containing abundant crinoids, corals and brachiopods, and dark gray to white chert.

Access is complicated. Take the Bedwell Lake trail from the S end of Buttle Lake in Strathcona Park. From Bedwell Lake you will need crampons and ice axe to cross the N Big Interior Glacier. The outcrop is only exposed from July through September, and is subject to snow in any month.

Mt Septimus to the E has a limestone showing that outcrops over a 300 m by 350 m area on the W side, on the S end of the Buttle Lake uplift. It is underlain by volcanic breccia, tuff and argillite, and to the NE by pillow basalts (092F 092).

78. MINFILE NUMBER: 092F 220   * Geology
NAME(S): CREAM 1, 3

Status:	Prospect		
NTS Map:	092F 05E	UTM Zone:	10
Latitude:	49 29 09	Northing:	5484350
Longitude:	125 32 06	Easting:	316400
Elevation:	1280 M		
Comments:	Located within a few hundred metres of the SW shores of Cream Lake. The site of considerable environmental protest in the 1980s when it was proposed to drain the lake and develop a mine at 1200 m in the park. Combine this trip with Big Interior Mountain (092F 090).		
Commodities:	Silver, gold, zinc, lead, copper		

MINERALS

Significant:	Arsenopyrite, sphalerite, galena, owyheeite, pyrite, tetrahedrite, pyrargyrite, chalcopyrite
Associated:	Quartz, siderite, calcite

The Cream vein has been exposed intermittently for at least 135 m along a strong lineament which is visible for about 1370 m. The vein strikes E and dips 65 degrees to 80 degrees N with its widths varying from 7 cm to 60 cm. It consists of quartz plus minor siderite and calcite lying within a strong gouge zone. Mineralization occurs as narrow bands or small masses in the quartz and consists of arsenopyrite, sphalerite, galena, owyheeite, pyrite, tetrahedrite, pyrargyrite and chalcopyrite.

The W side of Mt Septimus (on the E of Cream Lake) shows recrystallized limestone outcrops over a 280 m by 350 m area. It is underlain by volcanic breccia, tuff and argillite.

QUADRA ISLAND

Quadra Island is the most intensely mineralized and explored island in Johnstone Strait. Two distinct areas are of interest. Close to the Campbell River ferry, a dozen sites stretch NW from the N end of Gowland Harbour towards Morte Lake. Some 4 km N of Heriot Bay turn L onto Comox Dump Rd and a km later, turn left down towards the sea. At the shore turn up right into an unnamed valley. Both 079 and 080 are close to 1.2 km up the road.

Beyond Heriot Bay, a NNW-trending valley stretches from Open Bay to Granite Bay, showing over thirty recorded sites where the SW andesite formations contact the NE limestone. Quartz veining and copper sulfides are common. The Open Bay Epithermal site shows some unusual additional minerals. Take the road round Hyacinthe Bay from Heriot Bay village and follow the signs to Granite Bay.

79. MINFILE NUMBER: 092K 052 ** Minerals
   NAME(S): RADIUM

Status:	Showing		
NTS Map:	092K 03W	UTM Zone:	10
Latitude:	50 07 00	Northing:	5553843
Longitude:	125 16 00	Easting:	337950
Elevation:	120 m		
Comments:	Interesting and unusual minerals; useful tool is a geiger counter. Approach via Heriot Bay village and Comox Dump Rd.		
Commodities:	Vanadium, uranium, copper		
MINERALS			
Significant:	Carnotite, chalcocite		
Associated:	Quartz		

The Radium area is underlain by andesite and basalt. Amygdules are filled with chlorite, quartz, calcite and amphibole and locally chalcocite. Flows contain disseminated chalcocite and a fractured, thinly banded, black siliceous carbonaceous rock that carries vanadium values. Sparsely disseminated chalcocite and some malachite staining are also present within this black rock which is also cut by minute quartz veinlets. Carnotite occurs in fractures within the volcanic rocks. An analysis of a carnotite sample taken in 1932 gave 24% uranium and 21% vanadium oxide.

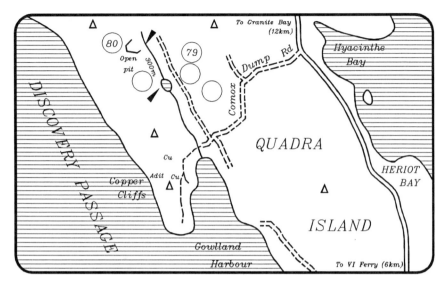

80. MINFILE NUMBER: 092K 071   ** Minerals
NAME(S): POMEROY 3,4, INGERSOLL

Status:	Past producer, open pit		
NTS Map:	092K 03W	UTM Zone:	10
Latitude:	50 07 05	Northing:	5554000
Longitude:	125 16 15	Easting:	337650
Elevation:	113 m		
Comments:	Open pit, 1 km up from N end of Gowlland Bay. Approach via Heriot Bay village and Comox Dump Rd.		
Commodities:	Copper, silver		

MINERALS

Significant:	Chalcocite, copper, chalcopyrite
Associated:	Quartz, calcite

The Pomeroy 4 comprises chalcocite mineralization controlled by strong fractures in amygdaloidal andesite flows. Malachite is prevalent as an oxidation product. The N trending fault separating the two zones contains high grade chalcocite mineralization, and minor native copper and chalcopyrite. A vein of quartz-calcite up to 38 cm wide and mineralized with chalcocite was previously explored.

The headland at the N end of Gowlland Harbour offers numerous outcroppings of chalcopyrite. An adit on the W slope overlooking Discovery Passage is located above the aptly named Copper Cliffs.

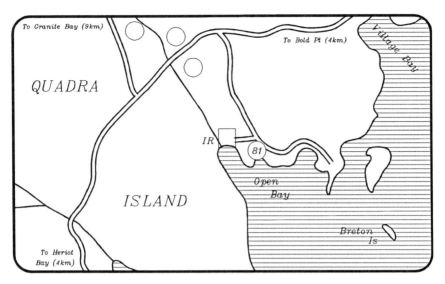

81. MINFILE NUMBER: 092K 005   * Minerals
NAME(S): OPEN BAY LIMESTONE

Status:	Showing		
NTS Map:	092K 03E	UTM Zone:	10
Latitude:	50 08 25	Northing:	5556350
Longitude:	125 12 27	Easting:	342250
Elevation:	91 m		
Comments:	Limestone interbedded with volcanic rocks occurs in a belt 1.2 km wide NW from Open Bay. Mostly on Open Bay Reserve.		
Commodities:	Limestone		

MINERALS

Significant: Calcite

The limestone is generally black and granular and emits a distinct odour of hydrogen sulphide when broken. Fine laminae of argillaceous impurities are distributed throughout the rock. Just a few hundred metres N, (Open Bay Epithermal) 092K 051 is a showing of unusual minerals.

Commodities: Mercury, arsenic, antimony

MINERALS

Significant: Cinnabar, stibnite
Associated: Pyrite, marcasite, pyrrhotite

The Open Bay Epithermal occurrence is located 400 m N of Open Bay. The epithermal zone occurs in strongly brecciated limestone. The zone contains cinnabar, stibnite and small amounts of pyrite, marcasite and pyrrhotite. Some areas are massive, fine-grained and black with disseminated stibnite and cinnabar on fracture plane surfaces. Other areas are light coloured and very porous with disseminated cinnabar.

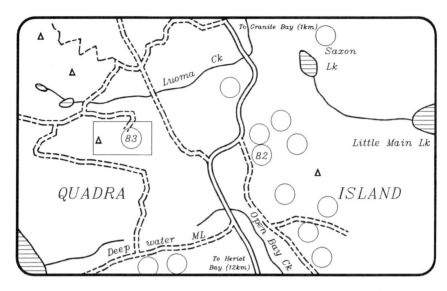

82. MINFILE NUMBER: 092K 015   ** Minerals
NAME(S): LUCKY JIM

Status:	Past producer, underground		
NTS Map:	092K 03W	UTM Zone:	10
Latitude:	50 12 20	Northing:	5563750
Longitude:	125 16 43	Easting:	337400
Elevation:	90 m		
Comments:	From the start of Granite Bay Rd, drive 5.2 km N. Where the road curves right and then sharp left, turn right (SW) onto gravel road for 300 m.		
Commodities:	Gold, silver, copper		

MINERALS

Significant:	Pyrrhotite, chalcopyrite, pyrite, marcasite, gold, sylvanite, telluride
Associated:	Quartz

The Lucky Jim deposit is situated 4 km SE of Granite Bay. The ore material follows a prominent line of faulting within the andesite but occurs along the limestone-andesite contact in the shaft area. The ore material consists almost entirely of pyrrhotite with some chalcopyrite, pyrite and marcasite. At other points along its strike this deposit includes more quartz, epidote, garnet and other silicates, and to the SE of the shaft a mass of magnetite is exposed. Free gold and sylvanite were also reported. The shaft was reported to be down 46m with ore still present near the bottom. Drifts are present at the 15 m and 30 m levels with drifts on the latter totalling some 67 m.

Two parallel zones of mineralization occur 90 m to the N and 90 m to the S of the Lucky Jim shaft. All ore deposits in the area occur in the vicinity of limestone.

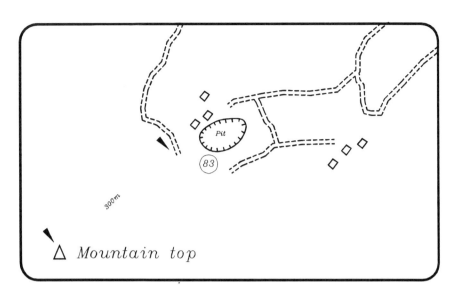

83. MINFILE NUMBER: 092K 060   * **Minerals**
     NAME(S): COPPER ROAD

Status:	Past producer, underground		
NTS Map:	092K 03W	UTM Zone:	10
Latitude:	50 12 26	Northing:	5564000
Longitude:	125 18 32	Easting	335250
Elevation:	457 m		

Comments: Located on the NE corner of the hill N of Mt Seymour. From S end of Granite Bay road, turn onto gravel road after 3.6 km. Road contours the hill for 1.5 km. Turn left at crossroad and climb for 200 m, bear left into valley, and turn up left (S) after 800 m. Track climbs steeply to hill plateau and site.

Commodities: Copper, silver, gold

MINERALS

Significant: Bornite, chalcopyrite, copper

Associated: Quartz, calcite

The Copper Road occurrence is underlain by dark green andesitic lavas. Amygdaloidal areas contain zeolite and epidote, and in one place hematite and chalcopyrite-filled amygdules. A shear up to 9 m wide and 1400 m long contains quartz, calcite, bornite, chalcopyrite, native copper and malachite.

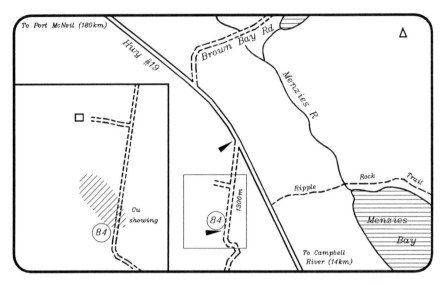

84. MINFILE NUMBER: 092K 068   ** Minerals
NAME(S): CHAL 4

Status:	Prospect		
NTS Map:	092K 03W	UTM Zone:	10
Latitude:	50 08 02	Northing:	5556088
Longitude:	125 24 50	Easting:	327487
Elevation:	152 m		
Comments:	Rare conditions to find volborthite (vanadium mineral).		
Commodities:	Copper, vanadium, iron, titanium		

MINERALS

Significant:	Chalcocite, volborthite
Alteration:	Malachite, azurite, brochantite

The Chal 4 is located approximately 16 km NW of Campbell River, W of Highway #19. The copper-vanadium minerals occur mainly within lenses of sedimentary rock. The seam is approximately 1 m thick at its widest point, strikes 315 degrees with a 45 degree NE dip and consists of black tuff-argillite overlain by fossiliferous limestone. The black tuff-argillite is heavily stained yellow, green and blue after chalcocite and volborthite. Malachite, azurite and bronchantite have also been identified. See also MINFILE 092K 066, a nearby showing.

CAMPBELL RIVER-PORT MCNEILL

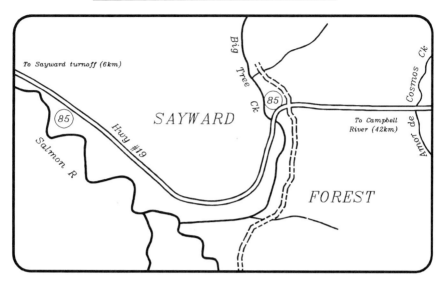

85. MINFILE NUMBER: 092K X02
NAME(S): BIG TREE CREEK

* * Gemstones

Status:	Prospect		
NTS Map:	092K 03W	UTM Zone:	10
Latitude:	50 15 00	Northing:	5565000
Longitude:	125 43 35	Easting:	30600
Elevation:	50 m		
Comments:	Dallasite placers in creek bed 33 km N of Bloedel/Menzies Bay on Highway #19.		
Commodities:	Dallasite breccia		

MINERALS

Significant:	Basalt breccia

Dallasite is reported in the bed of the creek close to the bridge at Big Tree Creek. See also at the highway bridge crossing the Salmon River, shortly before Sayward turn-off.

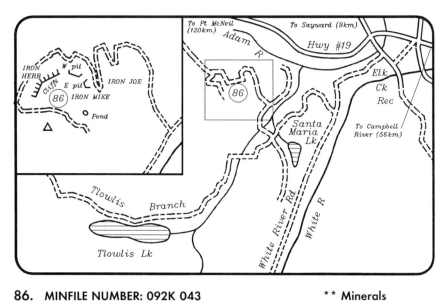

86. MINFILE NUMBER: 092K 043

** Minerals

NAME(S): IRON MIKE

Status:	Past producer, open pit		
NTS Map:	092K 05W	UTM Zone:	10
Latitude:	50 18 40	Northing:	5577229
Longitude:	125 58 20	Easting:	288377
Elevation:	400 m		
Comments:	Open pit, approximately 6 km S of Sayward.		
Commodities:	Iron, magnetite, copper		

MINERALS

Significant::	Magnetite
Associated:	Pyrite, chalcopyrite

The Iron Mike open pit is located approximately 6 km SSW of Sayward. The deposit is a garnet-epidote-magnetite skarn which occurs along the contact between an underlying greenstone and an overlying limestone. Mineralization is magnetite, essentially free of any impurities within the skarn. During production, mill feed grades averaged 45% iron with no contained impurities. Production began in 1965 and continued through to 1966, producing 82 thousand tonnes of iron from 168 thousand tonnes mined.

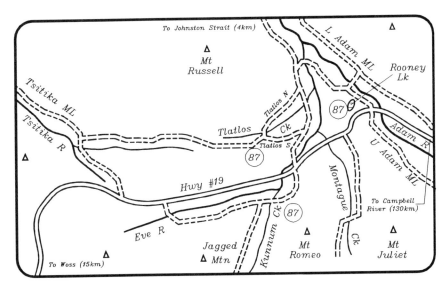

87. MINFILE NUMBER: 092L X01
NAME(S): ADAM AND EVE

** Gemstones

Status:	Showing		
NTS Map:	092L 8	UTM Zone:	10
Latitude:	50 20 00	Northing:	5579500
Longitude:	126 10 00	Easting:	70000
Elevation:	100 m		
Comments:	Quartz crystals intergrown with epidote.		
Commodities:	Quartz		

MINERALS

Significant: Quartz, epidote, dallasite breccia

Pockets of well formed quartz crystals as large as 15 cm have been found N and S of Highway #19 between Adam and Eve rivers. Early finds were on the logging roads S of the highway and Rooney Lake. More recent discoveries were recorded some 6 km up Tatlos Main where it leaves Main Line on the Eve River, and were made in new road cuttings.

BONANZA LAKE

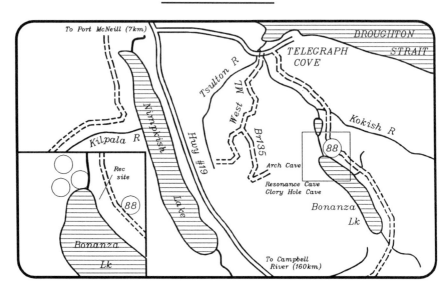

88. MINFILE NUMBER: 092L 339   * Geology
NAME(S): LEO D'OR

Status:	Developed prospect		
NTS Map:	092L 07W	UTM Zone:	09
Latitude:	50 23 44	Northing:	5584709
Longitude:	126 47 55	Easting:	656466
Elevation:	450 m		
Comments:	On the Leo d'Or claim, E of the N end of Bonanza Lake. Access from Telegraph Cove along the Kokish River. Follow signs to Bonanza Lake N Rec. Site.		
Commodities:	Marble		

MINERALS

Significant:	Marble
Comments:	Surface staining indicates iron impurities.

A band of limestone, up to 2.5 km wide, trends N along the E side of Bonanza River and Bonanza Lake for 6.3 km. On the property, the limestone has been recrystallized to marble by local intrusions. Discontinuous dykes of basalt, averaging 60 cm in width, are observed throughout the area. The marble varies in colour from very light gray to dark gray or almost black, to mottled gray and white. In some areas distinct black to light gray bands of marble, varying form several centimetres to several metres, occur. The grain size of the marble varies from fine to coarse grained. Light brown to light orange surface staining is caused by the oxidation of iron impurities.

ZEBALLOS

The area has a long history of gold mining, mostly from narrow quartz stringers which contain minute "colour." The main goldbearing formations are located on either side of the road linking Zeballos with Highway #19, about 5.5 km N of Zeballos, or 2.5 km S of the Nomash River turnoff (to Zeballos Lake and Rugged Mtn). The Zeballos River may be panned, or the two creeks on the SE that drain Spud and Gold Valleys are worth attention. On the NW side of the road, Blacksand and Lime creeks have similar potential. Access to the SE side of the valley is at the bridge 2 km N of Zeballos. Trails up Gold and Spuds Creeks are of unknown quality.

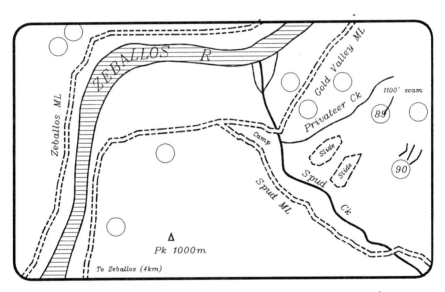

89. MINFILE NUMBER: 092L 008   ** Minerals
     NAME(S): PRIVATEER

Status:	Past producer, underground		
NTS Map:	092L 02W	UTM Zone:	09
Latitude:	50 01 50	Northing:	5544091
Longitude:	126 49 03	Easting:	656312
Elevation:	122 m		
Comments:	Main adit on Spud Creek, 0.5 km S of Zeballos River, 5.5 km NE of Zeballos.		
Commodities:	Gold, silver, lead, copper, zinc		

MINERALS

Significant::	Pyrite, sphalerite, galena, chalcopyrite, arsenopyrite, pyrrhotite
Associated:	Quartz, calcite

The Privateer mine lies in the Zeballos gold camp. Recorded production for the camp totals 9,465 kg of gold and 4,119 kg of silver from 652,000 tonnes of ore mined. Most production came from the Spud Valley deposits (092L 211 & 092L 013) and Privateer.

The Privateer mine, (1934-1975), consists of three roughly parallel main veins from which ore was produced, and more than twelve lesser, subsidiary veins. All veins follow shear zones. The veins are located in drag-folded andesitic tuff and hosts calc-silicate skarn.

The three veins from which most of the production was recorded contain alternating bands of quartz and sulfides. Locally comb textures and quartz-lined vugs up to 30 cm are present. Where sulfides are absent, variably altered wallrock inclusions are common. Coarse ankerite is often present. The productive parts of the veins contain abundant sulfides — pyrite, sphalerite, galena, chalcopyrite, arsenopyrite and pyrrhotite. Late calcite veinlets, overprinting the main veins, are often present.

90. MINFILE NUMBER: 092L 010   * Minerals
NAME(S): WHITE STAR

Status:	Past producer, underground		
NTS Map:	092L 02W	UTM Zone:	09
Latitude:	50 01 25	Northing:	5543341
Longitude:	126 48 25	Easting:	657091
Elevation:	284 m		
Comments:	No.3 adit is 150 m NE of Spud Creek, 1 km SE of Zeballos River, 5.5 km NE of Zeballos.		
Commodities:	Gold, silver, copper, lead, zinc		
MINERALS			
Significant:	Gold, chalcopyrite, galena, sphalerite, pyrite, arsenopyrite		
Associated:	Quartz		

The White Star mine lies in the Zeballos gold camp. Five veins are recognized at the White Star mine, all within quartz diorite intruded by NNE striking feldspar porphyry dykes. The veins lie 300 m E of the quartz diorite contact with calc-silicate altered tuffs of the Lower Jurassic Bonanza Group. The shear zones are up to 15 cm wide, the quartz veins contained in them are somewhat narrower. Diagonal gash veins, commonly filled with comb quartz, are common.

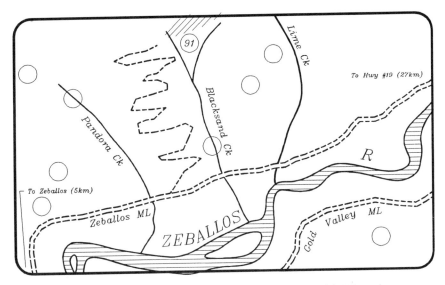

91. MINFILE NUMBER: 092L 028   *** * Minerals**
   NAME(S): FORD

Status:	Past producer, open pit, underground		
NTS Map:	092L 02W	UTM Zone:	09
Latitude:	50 02 55	Northing:	5546065
Longitude:	126 50 00	Easting:	655120
Elevation:	792 m		
Comments:	The centre of the ore body is 1.5 km N up Blacksand Creek from Zeballos River, 6.5 km N of Zeballos.		
Commodities:	Iron, magnetite crystals		

MINERALS
Significant: Magnetite

Mineralization consists of a 21 m body of massive magnetite that strikes NE and dips NW. It follows the limestone-tuff contact down dip, but crosses the stratigraphy where the contact becomes vertical at depth. A thin layer of pyrite is present locally at the magnetite-limestone contact. Pyroxene-epidote skarn, with only minor garnet, occurs as an irregular 31 m thick layer, 3 m above the magnetite.

Most of the magnetite is pure, massive and fine-grained; but it commonly occurs as octahedral grains up to 1.3 cm across. During 1962 and 1963 the deposit was mined by open pit methods. From 1963 to the end of production in 1969, underground methods were used. Between 1962 and 1969 the deposit produced over 1,100 tonnes of iron.

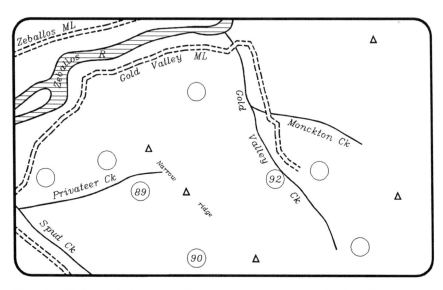

92. MINFILE NUMBER: 092L 015　　　　　　　* Minerals
NAME(S): LONE STAR

Status:	Past producer, open pit, underground		
NTS Map:	092L 02W	UTM Zone:	09
Latitude:	50 01 25	Northing:	5543370
Longitude:	128 47 35	Easting:	658086
Elevation:	450 m		
Comments:	Ore body centred on Gold Valley Creek above Zeballos River.		
Commodities:	Gold, silver, copper, lead, zinc		

MINERALS

Significant:	Gold, chalcopyrite, galena, sphalerite, pyrite, arsenopyrite
Associated:	Quartz

The Lone Star property lies in the Zeballos gold camp, an area underlain by basaltic to rhyolitic volcanic rocks. Conformably underlying these are limestones and limy clastics and basalts. Bedded rocks are predominantly NW striking, SW dipping

Recorded production for the camp totals 9465 kg gold and 4119 kg silver, from 650 thousand tonnes of ore mined. Most production came from the Spud Valley and Privateer deposits.

The occurrence, in which about eight veins are recognized, lies at the centre of a diorite stock covering about 460 m. The veins range from 1 – 4 cm in width. Mineralization consists of pyrite with lesser amounts of arsenopyrite, galena, sphalerite and locally chalcopyrite in a quartz gangue.

The main vein strikes 45 degrees and dips 80 degrees S has been traced horizontally over 200 and vertically for 100 m.

SYDNEY INLET

N of Tofino and S of Gold River, the outer (W) coast of Vancouver Island is accessible only by water. Spectacular fiords, misty mountains and complex topography make for a great trip. The area is overlain by the same Jurassic (and older) metamorphics that are found in Victoria and E of Ucluelet.

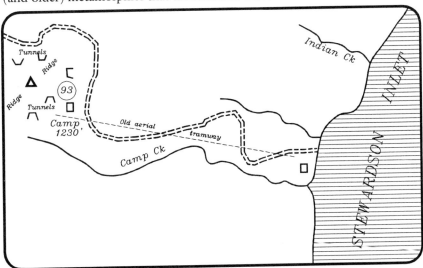

93. MINFILE NUMBER: 092E 011   ** Minerals
NAME(S): INDIAN CHIEF, MEPHISTOPHELES, BRUTUS

Status:	Past producer, underground		
NTS Map:	092E 08W	UTM Zone:	09
Latitude:	49 26 52	Northing:	5480493
Longitude:	126 18 38	Easting:	694938
Elevation:	502 m		
Comments:	Bonthorne adit 1.2 km W of Stewardson Inlet, off Sydney Inlet, due N of Hot Springs Cove. Not the sort of place for a Sunday field trip, but if you happen to be passing ...		
Commodities:	Copper, silver, gold, magnetite		

MINERALS

Significant::	Magnetite, bornite, chalcopyrite
Associated:	Pyrite

Mineralization consists of magnetite, bornite, chalcopyrite and pyrite. Higher copper grades occur in the fault zones and near the intrusive-limestone contacts. Magnetite occurs throughout the skarn. Production, from 1904 and 1938 was intermittent. Possible reserves are 1.3 million tonnes grading 1.6% copper.

The water at Hot Springs Cove surfaces at 51°C, flowing at 0.4 million l per day from a 15 cm shear zone. Salinity is 0.48 parts per thousand (sea water is 32 ppt).

MUCHALAT INLET

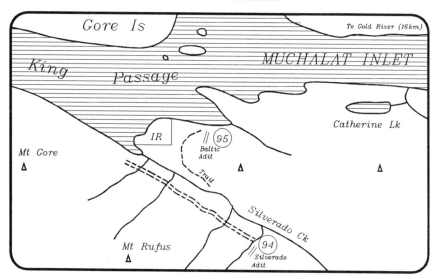

94. MINFILE NUMBER: 092E 017 ✱✱✱ Minerals
NAME(S): SILVERADO (L.1581), DANZIG, SHANNON, WYN, AM FR

Status:	Past producer, open pit		
NTS Map:	092E 09W	UTM Zone:	09
Latitude:	49 37 20	Northing:	5499754
Longitude:	126 21 40	Easting:	690594
Elevation:	91 m		
Comments:	Boat access only, opposite Gore Island. Location of adit is 1 km SE of the mouth of Silverado Creek on Kings Passage, Muchalat Inlet, 1 km S of Baltic (092E 026).		
Commodities:	Zinc, gold, silver, copper		

MINERALS

Significant:	Sphalerite, chalcopyrite, pyrrhotite
Associated:	Magnetite, quartz, diopside, calcite
Comments:	Chalcopyrite, pyrrhotite and magnetite in gangue of quartz-calcite and diopside.

The Silverado adit and workings explored a zone of discontinuous lenses of sphalerite that have partly replaced a 3 m wide bed of limestone along a greenstone contact. The contact strikes 330 degrees and dips 75 degrees W.

The zone of sphalerite lenses is traceable on the surface for over 100 m. As seen on the surface and in the drift, the lenses range in width from a few cm to 2 m and in length from 8 m to 30 m.

The mineralized lenses consist of sphalerite with small amounts of chalcopyrite, pyrrhotite and magnetite in a gangue comprised mainly of quartz, calcite and light-green diopside. Much of the mineralization is rhythmically banded with layers of sphalerite and gangue 0.2 cm to 2.5 cm thick. Between 1934 and 1938, 130 tonnes of ore from the Baltic and Silverado adits reportedly produced 5.5 kg of gold, 10 kg of silver and 87 kg of copper. Some 27 thousand tonnes of medium-grade tin remain.

95. MINFILE NUMBER: 092E 026   ** Minerals
NAME(S): BALTIC

Status:	Past producer, underground		
NTS Map:	092E 09W	UTM Zone:	09
Latitude:	49 37 53	Northing:	5500780
Longitude:	126 21 30	Easting:	690759
Elevation:	50 m		
Comments:	See Silverado for access. Location of adit is on King Passage, Muchalat Inlet, 1 km N of SILVERADO (092E 017).		
Commodities:	Gold, silver, copper		

MINERALS

Significant:	Pyrite, sphalerite, pyrrhotite, chalcopyrite
Comments:	Gold, silver associated with chalcopyrite.
Associated:	Quartz, magnetite
Alteration:	Mica, epidote, pyrite
Comments:	Alteration on vein margin.

Mineralization is believed to be related to the dykes, and occurs in nine veins. The No.1, 2 and 3 veins are linked by quartz stringers, and of this group only the No.1 vein is described. The vein is exposed in the first 42 m of the main adit. It is 10 cm to 30 cm wide and has been traced from the shore of Muchalat Inlet to an elevation of 76 m, a distance of 305 m. The vein strikes NNE and dips 70 degrees E. The vein consists of quartz, pyrite and sphalerite.

TLUPANA INLET

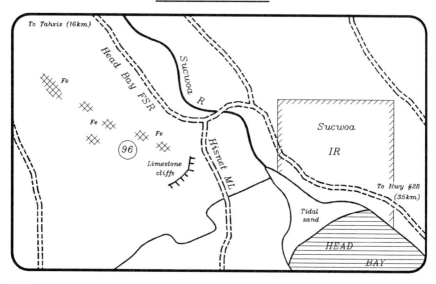

96. MINFILE NUMBER: 092E 001   * Geology
NAME(S): GLENGARRY, HEAD BAY, STORMONT

Status:	Past producer, open pit		
NTS Map:	092E 15E	UTM Zone:	09
Latitude:	49 48 28	Northing:	5520000
Longitude:	126 30 55	Easting:	678780
Elevation:	133 m		
Comments:	Access via the Gold River-Tahsis road (gravel). Located 1.8 km NW of Head Bay, W of Sucwoa River.		
Commodities:	Iron, magnetite		

MINERALS

Significant: Magnetite, pyrite, chalcopyrite

At the Glengarry magnetite occurrence, the NW striking Quatsino limestone dips about 45 degrees to the SW. Intruding the limestones to the S and E is a large granodiorite body and associated diorite dykes. The limestone strata have been recrystallized or altered to garnetite and many of the crosscutting dykes predate the skarn event.

Mineralization outcrops over an area of 550 m by 400 m as eleven or more pods of magnetite within garnet skarn. The pods range from 2 m to 12 m wide and are parallel to bedding, following roughly the margin of the intrusive contact in a NW direction. Chalcopyrite and pyrite are present only in small quantities. A sample of the magnetite assayed 57% iron.

TAHSIS

The town of Tahsis stands on a narrow strip of Quatsino Formation that is sandwiched between the NW trending Karmutsen Formation to the E and the younger (Jurassic) Bonanza Group to the W.

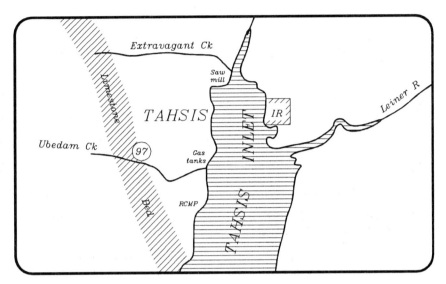

97. MINFILE NUMBER: 092E 010   **** Minerals**
NAME(S): GEO, STAR OF THE WEST

Status:	Prospect		
NTS Map:	092E 15E	UTM Zone:	09
Latitude:	49 55 05	Northing:	5531914
Longitude:	126 39 55	Easting:	667604
Elevation:	457 m		
Comments:	Main showing is at 460 m elevation on Ubedam Creek.		
Commodities:	Gold, silver, copper, lead, zinc, magnetite, iron, arsenic		

MINERALS

Significant:	Chalcopyrite, galena, sphalerite, magnetite, arsenopyrite, bornite
Associated:	Pyrite, pyrrhotite

Mineralization has been exposed by trenches over 50 m with a maximum width of 5 m, and consists mainly of lenses of chalcopyrite, magnetite, pyrite, pyrrhotite and local arsenopyrite (with associated gold).

Close by, a band of limestone (MINFILE 092E 070) up to 1.5 km wide extends NNW for 10 km. Outcrops display white to gray limestone (marble) that is dolomitic in places.

BROOKS PENINSULA

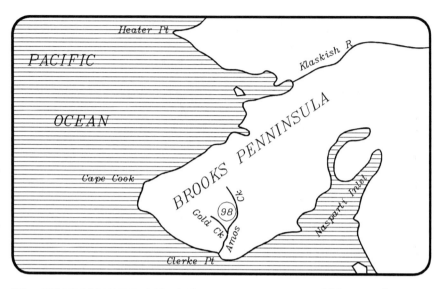

98. MINFILE NUMBER: 092L 248   *** Minerals
NAME(S): AMOS CREEK

Status:	Showing	Mining Division:	Alberni
NTS Map:	092L 04W	UTM Zone:	09
Latitude:	50 06 42	Northing:	5551501
Longitude:	127 48 48	Easting:	584849
Elevation:	61 m		
Comments:	3 km upstream from where Amos Creek empties into sea on SE corner of Brooks Peninsula.		
Commodities:	Gold		

MINERALS

Significant:	Gold
Comments:	Placer

The area is underlain by the West Coast Complex, comprised of a high grade crystalline, metamorphic complex derived from pre-Lower Mesozoic volcanic and sedimentary rocks. Coarse and fine placer gold was found over 5 km near the junction of Amos and Gold Creeks. Such remote access makes this site only worth visiting if you are passing in your kayak.

BENSON LAKE-ALICE LAKE

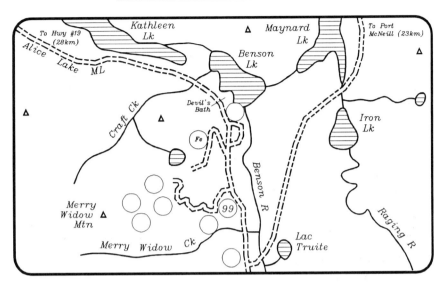

99. MINFILE NUMBER: 092L 091
NAME(S): BENSON LAKE, DRY HILL

*** **Minerals**

Status:	Past producer		
NTS Map:	092L 07W	UTM Zone:	09
Latitude:	50 16 18	Northing:	5584709
Longitude:	127 14 00	Easting:	656466
Elevation:	220 m		

Comments: Significant mineralization and exploration at the S end of Benson Lake and Merry Widow Mountain to the SW. About half way along the S side of Benson lake, the Devil's Bath is an interesting limestone sinkhole. Further caves occur in the limestone if you follow the Benson River S. Notices on the road to Atluck Lake identify the Vanishing River and (surprise!) the Reappearing River.

Commodities: Copper, magnetite

MINERALS
Significant: Chalcopyrite, magnetite, garnet
Comments: Well formed garnets

The S end of Benson Lake and the slopes on the W side have seen extensive exploration. Cobaltite and chalcopyrite occur with garnet alteration. Cobaltite is oxidized to erythrite. Skarn alteration minerals include epidote and serpentine. Silicas, pyrite and graphite present.

Three mine dumps are found along the Benson River between Merry Widow Creek and Benson Lake (two on W, one on E). A steep road climbing NW then switching SW cuts magnetite and sulphide bed about 40 cm thick, about 200 m beyond the bend.

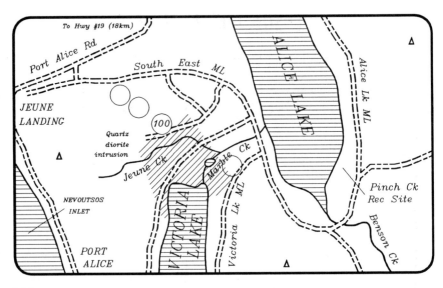

100. MINFILE NUMBER: 092L 314　　　　　　　* Minerals
NAME(S): BIG ZINC, CLANCY

Status:	Prospect		
NTS Map:	092L 07W	UTM Zone:	09
Latitude:	50 26 20	Northing:	5588385
Longitude:	127 25 58	Easting:	611293
Elevation:	250 m		
Comments:	Location is 2 km N of Victoria Lake and 2 km W of Alice Lake.		
Commodities:	Zinc		

MINERALS

Significant:	Sphalerite

A massive sphalerite occurrence is reported to occur in skarned rocks near the limestone-quartz diorite contact. Similar mineralization occurs about 3.2 km SE at the Peerless showing (Minfile Number 092L 057). About 45 hundred tonnes of ore grading between 4% and 5% zinc was reported to occur at surface exposures.

PORT HARDY AREA

Port Hardy can boast the very earliest serious mining on Vancouver Island. In 1835, natives reported finding coal in the area. In 1836 the Hudson's Bay Company ship "Beaver" was dispatched from Victoria to check these reports at Beaver Harbour and Suquash near today's Port Hardy. On the strength of a promising geological report, a hundred miners were sent out from England, arriving at Victoria in June of 1849, and moving on to the northern end of the island. It was quickly discovered that the coal was of poor quality (full of mudstone and shales), and the mining ceased the following year. Very fortuitously, in 1849 another native reported finding coal at Nanaimo (see sites 47 and 48), the miners moved from the north, and the rest is history ...

At Suquash, coal seams outcrop between Thomas Point at the S end of Beaver Harbour to Port McNeill — a distance of some 30 km. There is a shaft at shoreline level into a 1.3 m thick seam immediately SW of Single Tree Point.

Until 1996, Port Hardy boasted one of only three operational mines on Vancouver Island. BHP Minerals operated Island Copper for twenty seven years until the ore body ran out. The area is currently being reclaimed and replanted. Nevertheless, the area abounds in a multitude of interesting mineralogy.

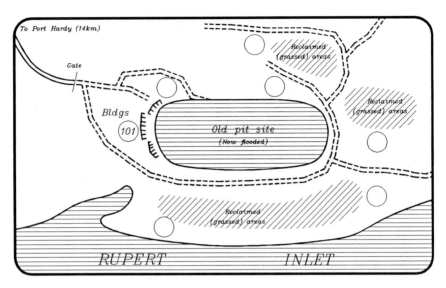

101. MINFILE NUMBER: 092L 158
NAME(S): ISLAND COPPER

* * * Minerals, Geology

Status:	Past producer, underground		
NTS Map:	092L11W	UTM Zone:	09
Latitude:	50 36 00	Northing:	5606238
Longitude:	127 28 24	Easting:	608045
Elevation:	76 m		
Comments:	Access is uncertain from Port Hardy. The beach offers an easy approach from Rupert Inlet over a 3 km length on the S side. A gated track from Frances Lake may be the best route from the N.		
Commodities:	Copper, molybdenum, silver, gold, rhenium, zinc, lead		

MINERALS

Significant:	Pyrite, chalcopyrite, molybdenite, bornite, sphalerite, galena
Associated:	Quartz, carbonate, dumortierite

The Island Copper deposit lies within brecciated tuff of the Bonanza Group pyroclastic sequence. These volcanic rocks are cut by a quartz feldspar porphyry dyke. Breccias with volcanic and intrusive fragments cap the dyke and occur along its margins. Brecciation is less intense a short distance outward from the porphyry and within about 60 m the dislocated breccia gives way to systems of intense fracturing (crackle breccia).

Although pyrite is the most abundant sulphide, chalcopyrite and molybdenite were the only sulfides recovered. Sphalerite and galena occur erratically in carbonate veinlets within and peripheral to the ore zone. Bornite has been observed in the ore zone in negligible quantities. Oxide minerals include magnetite, hematite and leucoxene.

Locally, brecciated hydrothermally altered andesitic to basaltic lapilli tuff and tuff breccia of the Bonanza Group are cut by a NW trending quartz-feldspar porphyry dyke swarm. At the W edge of the deposit the dyke is overlain by an intensely altered breccia zone up to 150 m wide and 1 km long. The zone is massive grayish-tan pyrophyllite, quartz and kaolin, speckled with bright-blue dumortierite. The zone appears brecciated and has a coarse, gritty texture.

The open pit operation went to 340 m below sea level, resulting in the lowest man-made point on earth. To prevent seepage from the inlet close by, engineers built a concrete slurry wall about 1 m thick, 2 km wide, and 350 m deep to act as a dam, between the hole and the shoreline. The mine closed late in 1995 and there has been extensive reclamation and landscaping by BHP.

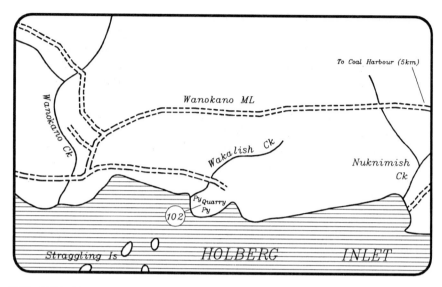

102. MINFILE NUMBER: 092L 150   * Minerals
NAME(S): HOLBERG, BETTY, RAINBOW

Status:	Developed prospect		
NTS Map:	092L12E	UTM Zone:	09
Latitude:	50 36 44	Northing:	5607300
Longitude:	127 40 55	Easting:	593250
Elevation:	10 M		
Comments:	Access is via the Wanokano Mainline W of Coal Harbour for 10 km until it crosses the Wanokano Creek and reaches Holberg Inlet. Take shore road back (E) 2 km.		
Commodities:	Silica		

MINERALS

Significant:	Silica
Associated:	Pyrite, fossils

On the claim there is a silica rich knoll in rhyolitic volcanics. Fragments up to 2.5 cm in diameter have been noted. The rocks are white to very light gray, hard and intensely fractured. Abundant pyrite occurs in irregular dark gray vesicular masses within the rhyolitic rocks on the W side of a small quarry.

Alteration consists chiefly of silicification with some kaolinization and the pyritization. The secondary silica is partly chalcedony and partly microcrystalline quartz. In 1965, 45 hundred tonnes of rhyolite were shipped to Vancouver by Lafarge Cement of North America Ltd.

Fossils are reported in the siltstones showing on the N side of Apple Bay, between Holberg and Coal Harbour. Poor quality coal deposits (similar to Suquash on the Johnstone Strait side of the island) were worked from 1883 to 1885 at Coal Harbour, but work did not continue.

SAN JOSEF

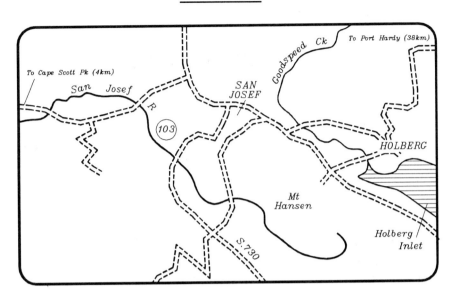

103. MINFILE NUMBER: 102I 012
NAME(S): REALGAR, ORP

** Minerals

Status:	Showing		
NTS Map:	102I 09E	UTM Zone:	09
Latitude:	50 39 00	Northing:	5611086
Longitude:	128 05 00	Easting:	564806
Elevation:	55 m		
Comments:	Located in San Josef River, 5.8 km W of Holberg.		
Commodities:	Mercury, arsenic, gemstones		

MINERALS

Significant:	Cinnabar, realgar, arsenopyrite
Associated:	Pyrite, amethyst, quartz, calcite

The Realgar occurrence is located at waterline on the E side of San Josef River, and is exposed only at low water. Locally, limestone is cut by a feldspar porphyry dyke of unknown age. Within about 4 m of the dyke the limestone is cut by irregular veins comprised of fine-grained chloritic skarn that range up to 10 cm in width. Cinnabar and realgar are abundant as thin veinlets and as fracture coatings in the limestone, and as accessory minerals in the skarned veins. Orpiment occurs as coatings on the oxidized mercury-arsenic minerals. Traces of amethyst occur in veinlets. Arsenopyrite and pyrite occur as disseminations in the limestone near the dyke and within thin, vuggy quartz-calcite veinlets cutting both the limestone and the dyke. High mercury values with coincident arsenic and molybdenite values are reported in soils over a wide area centered on the showing.

The exact location of this unusual deposit has proven elusive. Best access is to put in a boat at the ramp on the San Josef River where the Port Hardy — Cape Scott Provincial Park road bridges the river the first time.

TIPS, TRICKS AND TRIVIA

SCIENCE PROJECT

Magnetite was known to the Norsemen as "lodestone." It was their first compass, which they used to navigate the Atlantic Ocean many centuries ago. Find a small piece of black magnetite. The difference between magnetite and hematite is that, although both are usually black and heavy and will let a magnet stick to them, only magnetite will affect a compass if you bring one close to the sample. Hematite does not make the needle "wobble." Place the magnetite on a floating block of wood in a bowl of water. This makes a frictionless bearing. Allow the block to rotate slowly until it settles in one orientation. Using a conventional compass, or the sun at midday, or the Pole Star at night, determine magnetic north and place a dot of paint on the magnetite to identify its N pointing pole. Congratulations: you have just created a nordic compass. Now, try using it at sea!

PROSPECTING TIP

Although rhodonite can be seen in many tourist stores on Vancouver Island in the form of bright pink carvings (bears, eagles, etc), in its natural state it oxidizes to manganese oxide, which is black and uninteresting. When hunting rhodonite, therefore, look for black rocks. Use a hammer to test for pink.

PROSPECTING TIP

When panning for gold, remember the metal is very dense (about 5 times heavier than most rocks), so it settles readily into deep cracks. Look for black sand (magnetite) which is often associated with it. Test for black sand with a small magnet. Also look for tiny pink/red garnets in the bottom of the pan. If you're finding them, then you're panning OK, and will find gold if there's any about.

PROSPECTING TIP

When looking for sulphides (pyrite, chalcopyrite, etc), use your nose — it's better than your eyes. Crack a sample to produce new surfaces, then sniff. If you smell rotten eggs (sulphur), you've found a sulphide!

SCIENCE TRIVIA

Hematite (or haematite) is a dark iron oxide that can be confused with magnetite, except it is not magnetic, and when streaked along a rough surface (like concrete) it leaves a brown/red line (rust) behind it. This curious colour discrepancy was first noted by the ancient Greeks, who named the mineral haema-lithos (literally: blood-stone) from which we derive the modern name hema-tite.

SCIENCE TRIVIA

Geologists use the Mohs Scale of Hardness to compare the 'scratchability' of one mineral to another. While every school child knows the hardest thing in the world is a diamond (well, almost the hardest thing) with H=10, few know that talc is one of the softest, with a hardness of 1. Talc is ground up and sold as talcum or baby powder.

SCIENCE PROJECT

Find a small piece of limestone (about the size of a golf ball). Place in the bottom of a narrow glass jar, and just cover it with white grape vinegar (not apple). Notice the bubbles. Leave in a warm spot. As the vinegar evaporates over a week, white crystals of aragonite (a form of calcite) grow on the rock and glass. When dry, do it all over again, only this time, add food colouring!

Explanation: limestone is an alkali (base); vinegar is a mild acid. As the water molecules evaporate, the free carbonate ions re-align, forming feathery crystals of calcium carbonate.

SOME FAVOURITE SITES

3	Island View Beach	agate
14	Muir Creek	fossils
15	Sunro	sulphides
17	Sombrio Placers	gold
21-24	Leechtown	gold
27	Peg	pegmatites
29	Hollings	rhodonite
30	Mesabi	jasper
32	Mt Sicker	sulphides
33	Hill 60	rhodonite
36	Blue Grouse	copper
37	Meade Creek	gold
41	Holyoak Lake	flowerstone porphery
44-46	Chemainus Valley	jasper
50	Main	agate
53	McQuillan	jasper
58	Brynnor	hematite, cave marble
59	Florencia Bay	gold
68	Upper Qualicum River	dallasite
69-70	Hornby Island	fossils
71	Mt Washington Copper	chalcopyrite
82	Lucky Jim	copper
99	Benson Lake	garnets
101	Island Copper	dumortierite

SITES BY PRINCIPAL MINERALS

Agate	12, 50,
Cinnabar	62, 103
Coal	47A, 47B
Copper	4, 15, 28, 31, 32, 35, 36, 56, 61, 64, 71, 76, 80, 82, 83, 93, 97, 99, 101
Dallasite	5, 10, 65, 66, 68, 85
Dumortierite	101
Flowerstone	13, 41, 72
Fossils	14, 16, 69, 70
Garnet	99
Gold	11, 17, 18, 19, 22, 23, 26, 37, 59, 89, 90, 92, 95, 98
Iron	9, 20, 46, 58, 73, 74, 86, 91, 96
Jasper	3, 30, 39, 44, 45, 46, 51, 53
Limestone	2, 6, 7, 8, 20, 55, 60, 67, 77, 81, 88
Pegmatite	27
Quartz	51, 52, 87
Rhodonite	29, 33, 34, 38, 40
Silica	102
Silver	78
Sphalerite	94, 100
Staurolite	25
Talc	21, 24
Tourmaline	25, 27
Vanadium	79, 84

CHAPTER 10: ADDRESSES

Reference Libraries:

Victoria

Ministry of Employment & Investment Library
Ground Fl., 1810 Blanshard Street
PO Box 9321, Station Provincial Government
Victoria, BC V8W 9N3
phone: 250- 952-0583, fax: 250-952-0581
University of Victoria Library
PO Box 1700
Victoria, BC V8W 2Y2

Vancouver

Geological Survey of Canada
Suite 1500, 605 Robson Street
Vancouver, BC V6B 5J3
phone: 604-666-3812, fax: 604-666-7186
BC Geological Survey
301 – 865 Homer Street
Vancouver, BC V6Z 2G3
phone: 604-660-2700

Ottawa

Geological Survey of Canada
601 Booth Street,
Ottawa, ON K1A 0E8
phone: 613-995-0947

Maps and Reports:

Victoria

Ministry of Employment & Investment (was Energy,
Mines & Petroleum Resources), Geological Survey Branch
5th Floor, 1810 Blanshard Street,
Victoria, BC V8V 1X4
fax: 250-952-0381
Crown Publications Inc
521 Fort Street,
Victoria, BC V8W 1E7
phone: 250-386-4636, fax: 250-386-0221

Vancouver

Geological Survey of Canada, Map & Publication Sales
Suite 1500, 605 Robson Street
Vancouver, BC V6B 5J3
phone: 604-666-3812, fax: 604-666-7186

Ottawa

Geological Survey of Canada
601 Booth Street
Ottawa, ON K1A 0E8
phone: 613-995-0947

Claim Maps and Free Miner Certificates:
To locate the office serving your community, contact BC Enquiry toll free at
1-800-663-7867.

Gold Commissioners' Offices:
(Some of these offices have a mines inspector, but not a resident geologist.)
Alberni Mining Div., 4515 Elizabeth St., Port Alberni, BC V9Y 6L5 (250) 724-9204
Atlin Mining Div., Third Street, Box 100, Atlin, BC V0W 1A0 (250) 651-7595
Cariboo Mining Div., 102 – 350 Barlow St., Quesnel, BC V2J 2C1 (250) 992-4301
Clinton Mining Div., 1423 Cariboo Highway, Clinton, BC V0K 1K0
    (250) 459-2268
Fort Steele Mining Div., 102 – 11th St., Cranbrook, BC V1C 2P2 (250) 426-1211
Golden Mining Div., 606 – 6th St. N., Box 39, Golden, BC V0A 1H0
    (250) 344-7550
Greenwood Mining Div., 524 Central Ave, Grand Forks, BC V0H 1H0
    (250) 442-5444
Kamloops Mining Div., 250 – 455 Columbia St., Kamloops, BC V2C 6K4
    (250) 828-4540
Liard Mining Div., 302 – 865 Hornby St., Vancouver, BC V6Z 2C5 (604) 660-2672
Lillooet Mining Div., 615 Main St., Bag 700, Lillooet, BC V0K 1V0 (250) 256-7548
Nanaimo Mining Div., 13 Victoria Cres., Nanaimo, BC V9R 5B9 (250) 755-2200
Nelson Mining Div., 310 Ward St., Nelson, BC V1L 5S4 (250) 354-6104
New Westminster Mining Div., 100 – 635 Columbia St., New Westminster, BC
    V3M 1A7 (604) 660-8666
Nicola Mining Div., 1840 Nicola Ave., Box 4400, Merritt, BC V0K 2B0
    (250) 378-9343
Omineca Mining Div., 3793 Alfred St., Box 5000, Smithers, BC V0J 2N0
    (250) 847-7207
Osoyoos Mining Div., 112 – 100 Main St., Penticton, BC V2A 5A5 (250) 492-1211
Revelstoke Mining Div., 1100 West 2nd St., Box 380, Revelstoke, BC V0E 2S0
    (250) 837-7636
Similkameen Mining Div., 151 Vermilion Ave., Box 9, Princeton, BC V0W 1W0
    (250) 295-6957
Skeena Mining Div., 100 Market Place, Prince Rupert, BC V8J 1B7 (250) 627-0415
Slocan Mining Div., Box 580, Kaslo, BC V0G 1M0 (250) 353-2219
Trail Creek Mining Div., 2888 Columbia Ave., Rossland, BC V0G 1Y0
    (250) 326-7324
Vancouver Mining Div., 302 – 865 Hornby St., Vancouver, BC V6Z 2C5
    (604) 660-2672
Vernon Mining Div., 102 - 3001 27th Str., Vernon, BC V1T 4W5 (250) 549-5511
Victoria Mining Div., 3rd Fl., 1810 Blanshard St., Victoria, V8W 9N3
    (250) 952-0542

Mining & Exploration Information:
Regional Geologist's Offices, Ministry of Employment & Investment:
Cranbrook: 1113 Baker Street, V1C 1A7 (250) 426-1557
Fernie: Bag 1000, V0B 1M0 (250) 423-6884
Kamloops: #200, 2985 Airport Drive, V2B 7W8 (250) 828-4566
Nanaimo: 13 Victoria Cres., V9R 5B9 (250) 751-7240 (note new phone number)
Prince George: 1652 Quinn Street, V2N 1X3 (250) 565-6125
Smithers: Bag 5000, V0J 2N0 (250) 847-7391
Vancouver: Rm 301, 865 Hornby St., V6Z 2G3 (604) 660-0223

Mining Museums:

Please note that, apart from the Britannia Beach Musuem, all 604 area codes have changed to 250.

Barkerville, Barkerville Historic Town, at the end of Hwy 26. (604-994-3332)

Britannia Beach, British Columbia Museum of Mining, on Hwy 99, 52 km from Vancouver. (604-688-8735)

Cumberland, Cumberland Museum, off Island Hwy near Comox. (250-336-2445)

Rossland, Rossland Mining Museum, junction Hwys 3B/22. (250-362-7722)

Silverton, Silverton Outdoor Mining Exhibit, on Hwy 6, near Silverton Village. (604-358-7788)

Woodberry Creek, Woodberry Mining Museum, Hwy 31, 2 km north of Ainsworth Hot Springs. (604-353-2592)

Mineral displays in BC:

Abbotsford Museum, 2313 Ware St., Abbotsford *(Mining exhibit)*

Ashcroft Museum, 404 Brink St., Ashcroft *(Mining & mineralogical displays)*

Atlin Museum, 3rd/Trainor Sts., P.O.Box 111, Atlin *(Gold rush material)*

Courtney & District Museum, 360 Cliffe Ave., Box 3128, Courtney *(Fossils)*

Dawson Creek Museum, 13 St/Alaska Ave., Dawson Creek *(Fossils, mammoth tusk)*

Fernie Museum, 502 – 5th Avenue, Box 1527, Fernie *(Mining gear (coal) and history)*

Peace Island Park Museum, Taylor, near Fort St John *(Fossils, rocks)*

Boundary Museum, Box 17, Grand Forks *(Mining equipment)*

Fort Steele Heritage Town, 16 km N Cranbrooke, Hwys 93/95 *(Mining exhibit)*

Grand Forks Boundary Museum, Hwy 3, 7370 – 5th St. *(Mining exhibit)*

Greenwood Museum, 214 South Copper St., Greenwood *(Mining exhibit)*

Hope Museum, 919 Water Str., Hope *(Mining exhibit)*

Hudson's Hope Museum, Box 98, Hudson's Hope *(Fossils, dinosaur and gold panning history)*

Kaatza Museum, Lake Cowichan, Vancouver Island *(Mining exhibit)*

Kamloops Museum, 207 Seymour St., Kamloops *(Local geology)*

Kelowna Centennial Museum, 470 Queensway, Kelowna *(Mining exhibit)*

Kettle River Museum, Hwy 3 (CPR Station), Midway *(Mining Exhibit)*

Kimberley Heritage Museum, 105 Spokane St., Kimberley *(Mining Exhibit)*

Kitimat Smelter, Kitimat *(Public tours daily)*

Lillooet Museum, Box 441, Lillooet *(Mining & gold rush history)*

Mission Museum, 33201 Second Ave., Mission *(Mineral collection)*

Nanaimo Centennial Museum, 100 Cameron Rd., Nanaimo *(Coal mining history, simulated mine tunnel)*

Nelson Centennial Museum, 402 Anderson St., Nelson *(Geology & mining history)*

Nicola Valley Museum, 2202 Jackson Ave., Merritt *(Mining exhibit)*

Osoyoos Museum, Box 791, Osoyoos *(Mining samples, fossils)*

R.N. Atkinson Museum, 785 Main St., Penticton *(Mineralogy and natural history displays)*

Powell River Historical Museum, Box 42, Powell River *(International sand collection)*

Fort George Regional Museum, Box 1779, Prince George *(Mineral displays)*

Museum of Northern BC, Box 669, Prince Rupert *(Rock collection, mining history)*

Quesnel Historical Museum, le Bourdais Park, 405 Barlow Ave., Quesnel *(Rock & mining displays)*

Rossland Historical Museum, Box 26, Hwys 22 & 38, Rossland *(Underground tours of LeRoy Mine)*

Sandon Museum, Box 303, Sandon *(Mining history)*
Bulkley Valley Historical & Museum Society, Box 2615, Smithers *(Some fossil material)*
Wells Historical Society, Box 244, Wells *(Mining history)*
Williams Lake Museum, Williams Lake *(Mining history)*
Yale Museum, 31179 Douglas St., Yale *(Mining exhibit)*
Mineral World, 9891 Seaport Pl., Sidney, BC, ph: (250) 655-4367, fax: (250) 656-2350 *(Mineral displays)*
Science World, 1455 Quebec Street, Vancouver, BC, ph: (604) 268-6363, fax: (604) 682.2923 *(Mineral displays)*
BC & Yukon Chamber of Mines, 860-1066 West Hastings St., Vancouver *(Small museum)*
Royal BC Provincial Museum, 601 Belleville, Victoria *(Some displays)*
BC Provincial Archives, Parliament Buildings, Victoria *(Manuscripts, historical documents)*
BC Ministry of Employment & Investment Library, 1st Fl., Jack Davis Building, 1810 Blanshard St., Victoria *(Rock & mineral displays)*

Rock collecting magazines:
Cab & Crystal, 7 Elizabeth St N., Unit 406, Mississauga, ON L5G 2Y8
Rock & Gem, 4880 Market St., Ventura, CA 93003-7783
Lapidary Journal, Suite 201 60 Chestnut Avenue, Devon, PA 19333-1312
Rock & Minerals, 1319 Eighteenth St. NW, Washington DC 20036-1802

Vancouver Island Clubs & Associations:
Most clubs change their president, address and phone number every few years. For the latest information, contact your nearest Visitor Information service, Chamber of Commerce or Municipality, or write to the Lapidary Rock & Mineral Society of BC, c/o 13515 – 1112 Ave., Surrey, BC V3R 2E9

Publications of the Gem & Mineral Foundation of Canada are available from affiliated clubs and societies, or from the Foundation's Secretary at 3492 Dundas Street, Vancouver, BC V5K 1R8.

Vancouver Island clubs:
Victoria Lapidary & Mineral Society, Victoria
Vancouver Island Faceters Club, Victoria
Cowichan Valley Rockhound Club, Duncan
Parksville & District Rock & Gem Club, Parksville
Alberni Valley Rock & Gem Club, Port Alberni
Courtney Gem & Mineral Club, Courtney
Ripple Rock Gem & Mineral Club, Campbell River
B.C. Paleontological Alliance
Vancouver Island Paleontological Museum Society
Vancouver Island Paleontological Society
Victoria Paleontological Society

Commercial rock & mineral outlets:
Mineral World & Scratch Patch, 9891 Seaport Place, Sidney, BC (250) 655-4367
The Rockhound Shop, 777 Cloverdale St., Victoria, BC (250) 475-2080
Missing Link Fossils, 833 Poplar Way, Qualicum Beach, BC (250) 752-3979

CHAPTER 11: GLOSSARY OF TERMS

Acicular ('a-sick-u-lar') – needle shaped

Adit – horizontal or inclined passage entering a mine

Alluvial – sand, silt or gravel deposited by water

Amorphous ('a-more-fuss') – having no crystal structure

Amygdaloidal ('a-mig-de-loy-dal') – small, round gas cavities in rock

Anticline – strata folded like the letter 'A'

Asteriated – showing a star-like pattern

Bar – sand or gravel bed in river

Basal – cleavage parallel to base of the crystal

Batholith – igneous rock formed deep underground and of huge size

Bladed – elongated, flat and thin

Bloom – decomposed ore by surface oxidation

Botryoidal ('bot-ree-oi-dahl') – rounded masses, resembling a bunch of grapes

Brecciated ('bree-she-ated') – material made of sharp-edged rock fragments cemented together

Cabachon ('cab-o-shon' or 'cab') – an oval stone with convex face, popular for brooches, etc

Carat – unit of weight = one fifth of a gram, or 200 milligrams

Chatoyant ('shat-oy-ant') – reflecting light in a streak, like a cat's eye

Cleavage – tendency of a crystal to split along a face or plane

Columnar – like a column

Conchoidal ('con-koy-dahl') – clam shell-shaped, cupped surface

Concretion ('con-cree-shun') - lump of dissimilar rock in sedimentary matrix

Coulee ('coo-lee') – steep-sided, small valley

Craze – tendency for gemstone to develop tiny cracks

Cryptocrystalline ('kripto-kristal-een') – crystals so small as to be invisible to the naked eye

Crust – outer layer of earth's surface

Crystal – regular, repeating arrangement of atoms, often showing symmetrical planar faces

Dendritic – branching or fern-like shape of one mineral crystallizing inside another

Dip – angle from the horizontal (0) to vertical (90) that ore body/strata makes with the ground

Drusy – surface covered in small crystals

Dyke – vertical body of rock cutting through major rock matrix

Element – single substance having all the same atoms; i.e. silver, carbon

Escarpment (scarp) – sharp rise in land, often a cliff

Extrusive – igneous rock that solidifies on the surface

Facet – flat surface polished onto a gemstone

Fault – displacement of rocks along a fracture zone

Fibrous – threadlike or needlelike crystals

Fire – gemstone's brilliance

Fissure – crack in rock; may be empty or filled

Float – rock fragments found on surface some distance from outcrop

Foliated – capable of being separated into thin sheets, like mica

Fracture – the texture or shape of a broken surface

Gemstone – any stone considered precious or semi-precious when cut or polished

Geode – ('gee-ode') a hollow or crystal/chalcedony-filled nodule

Glacial drift – material carried by advancing ice

Habit – characteristic manner or occurrence of a mineral

Hackly – a jagged fracture

Hydrous – ('hi-druss') containing chemically bonded water

Inclusion – foreign material in a mineral

Intrusive – rock that has pushed into pre-existing matrix or rock, solidifying below the surface

Lamellar – parallel arrangement of platy crystals

Lode – economic concentration of mineral or ore

Loess ('low-ess') – silt deposited by the wind

Lustre – appearance of light on a fresh surface

Magma – molten rock or lava from deep within the earth

Massive – mineral formed in large, homogenous mass

Matrix – the rock or mineral mass

Metamorphic – change in mineral or crystal due to heat, pressure or chemical change

Mineral – naturally occuring substance with defined chemical composition

Moraine – stones and boulders scraped into piles by a glacier's edges

Mucking – removal of waste material, usually during mining operations

Nodule – a ball-like mineral mass

Opaque – material through which light cannot pass

Orbicular – having circular patterns

Outcrop – exposed rock on earth's surface

Overburden – rock or soil to be removed before reaching economic ore

Pan – wash gravels for gold and other heavy minerals

Percolating – filtering action of underground water or mineral solutions through the matrix

Petrify – reformed as a stone

Placer – stream separation of heavy minerals

Pitch – angle of dip or inclination of a mineral bed or vein, at right angles to its strike

Pluton – igneous intrusive that doesn't reach the surface

Primary stone – one that has not been changed by natural forces since its formation.

Prismatic – splitting light into its component colours and reflecting some or all of same

Pseudomorph ('sue-doe-morf') – where one mineral replaces another, keeping the previous shape

Refraction – bending of light as it passes from one medium (air) into another (gemstone)

Reniform – kidney-shaped form

Rough – uncut gem material

Sagenitic ('saj-e-nit-ick') – having needle-like crystals of a foreign mineral

Schiller – lustre or iridescence due to internal reflection of light

Seam – mineral mass in a crack or between strata

Secondary stone – natural alteration of existing rock or mineral

Sectile – capable of being cut into shavings, like wax

Sediment – rock or mineral produced by weathering or erosion, and then deposited

Shaft – vertical or sloping mine entrance

Silicify – change into chalcedony or opal

Sintering – melting an ore and removing the silica-rich liquids

Smelt – melt a rock, ore or mineral to remove some fraction

Stratum (plural 'strata') – single, distinct sedimentary layer

Strike – direction ore body makes compared to north (000-360 degrees)

Suite – group of minerals or rocks occuring together

Syncline – folded strata like the letter 'U'

Tabular – shaped like a table

Talus – rock debris at foot of cliff or slope

Translucent – allows light, but not detail, to pass through

Transparent – allows both light and detail to pass through

Tumble – method of polishing rocks in a rotary drum

Twin – symmetrical intergrowth of two or more crystals of the same species

Vein – thin sheet or stringer of mineral deposit

Vesicles – spherical or oval cavities formed in lava by gas bubbles

Vitreous – glass-like lustre

Vug – small cavity in rock, often containing a mineral

Approximate field conversions:

Metric to Imperial:

Length:	1 centimetre = 0.4 inches = width of fingernail
	1 metre = 39.2 inches = long yard (three paces)
	1 kilometre = 0.63 miles
Weight:	1 gram = 5 carats = 1/28th of troy ounce
	1 kilogram = 2.2 pound = weight of rock hammer

Imperial to Metric:

Length:	1 inch = 2.54 centimetre = distance between 2 knuckle lines
	1 foot = 30 centimetre = 1 adult shoe length
	1 yard = 0.92 metre = 3 paces
Weight:	1 ounce = 28.4 grams
	1 pound = 0.46 kilogram = 460 gram

Common chemical elements of mineralogy and their symbols

Aluminum	Al	Molybdenum	Mo
Antimony	Sb	Nickel	Ni
Arsenic	As	Nitrogen	N
Barium	Ba	Oxygen	O
Beryllium	Be	Phosphorus	P
Bismuth	Bi	Platinum	Pt
Boron	B	Potassium	K
Cadmium	Cd	Radium	Ra
Calcium	Ca	Selenium	Se
Carbon	C	Silicon	Si
Chlorine	Cl	Silver	Ag
Chromium	Cr	Sodium	Na
Cobalt	Co	Strontium	Sr
Copper	Cu	Sulfur	S
Fluorine	F	Tellerium	Te
Gold	Au	Thorium	Th
Hydrogen	H	Tin	Sn
Iron	Fe	Titanium	Ti
Lead	Pb	Tungsten	W
Lithium	Li	Uranium	U
Magnesium	Mg	Vanadium	V
Mangenese	Mn	Zinc	Zn
Mercury	Hg	Zirconium	Zr

CHAPTER 12: SELECTED READING

A field guide to rocks and minerals. *F.H. Pough. Houghton Mifflin Co., Boston, MA, 333pp, 1953.*

A list of Canadian Mineral occurences. *R.A.A. Johnston. Govt. Printing Bureau, Dept Mines, Ottawa, ON, 1915.*

An introduction to prospecting. *BC Ministry Energy, Mines & Petroleum Resources. Paper 1986-4, Crown Publications, Victoria, BC.*

Asbestos in British Columbia. *A.M. Richmond. BC Dept of Mines, Victoria, BC, 1932.*

Backroad & outdoor recreation MAPBOOK of Vancouver Island. *R. & W. Mussio. Mussio Ventures Ltd., Surrey, BC, 80pp, 1994.*

BC gem trails, 4th edition. *Howard Pearson. Private printing, 1973.*

British Columbia — the pioneer years, vols I, II, III. *T.W. Paterson. Stagecoach Publishing, Langley, BC, 128pp, 1980.*

Canadian Gem Stones. *Ron Purvis. Lillooet, BC, 29pp, 1962.*

Collecting minerals: a handbook for the amateur. *Bill Ince. McClelland & Stewart, Toronto, ON, 1980.*

Dredging for gold, the gold divers' handbook. *Matt Thornton. Keene Industries, Northridge, CA, 238pp, 1975.*

Exploring Minerals & Crystals. *R.I. Gait. McGraw-Hill Ryerson Ltd., Toronto, ON, 119pp, 1972.*

Famous mineral localities of Canada. *Joel D. Grice. Fitzhenry & Whiteside Ltd., Markham, ON, 190pp, 1989.*

Fur, gold & opals in the Thomson River Valleys. *M. Shewchuck. Hancock House, Surrey, BC, 128pp, 1975.*

Gems and minerals in color. *R. Metz. Hippocrene Books, Inc., NY, NY, 255pp, 1974.*

Gems and precious stones of North America. *George F. Kunz. Dover Publications Inc., NY, NY, 1892, republ. 1968.*

Gemstones of North America. *John Sinkankas. D. Van Nostrand Co. Inc., Princeton, NJ, 675pp, 1959.*

Geology of Canadian beryllium deposits. *R. Mulligan. Dept Energy, Mines & Resources, Ottawa, ON, 1968.*

Geology of Strathcona Provincial Park. *BC Ministry of Energy, Mines & Petroleum Resources, Geological Survey Branch. Information Circular 1995-7, Victoria, BC.*

George's guide to claimstaking in British Columbia. *Ministry of Energy, Mines & Petroleum Resources. Mineral Titles Branch, Victoria, BC, 1989.*

Glacier and Mount Revelstoke National Park, British Columbia; where rivers are born. *D.M. Baird. Misc. Report No.11, GSC Ottawa, ON, 104pp, 1965.*

Gold panner's manual. *Garnet Basque. Stagecoach Publishing, Langley, BC, 88pp, 1975.*

Guide to BC rocks and gems. *Western Homes & Living, n36, October 1961*

Guide to rocks and minerals of the northwest. *Stan & Chris Leaming. Hancock House, N.Vancouver, BC, 1980, reprinted 1992.*

Heather's amazing discovery. *D. Griffiths. Courtney & District Museum, Courtney, BC, 33pp, 1995.*

Historic treasures and lost mines. *N.L. Barlee. Canada West Publications, Summerland, BC, 128pp, 1978.*

How Old Is That Mountain? *C.J. Yorath. Orca Book Publishers, Victoria, BC, 176 pp, 1997.*

Jade in Canada. *S.F. Leaming. GSC Paper 78-19, GSC, Ottawa, ON, 59pp, 1977.*

Jade, stone of heaven. *National Geographic magazine, vol.172, n3, 33pp, Sept 1987.*

Kootenay National Park, British Columbia; wild mountains and great valleys. *D.M. Baird. Misc. Report No.9, GSC Ottawa, ON, 94pp, 1964.*

Mineralogy for students. *M.H. Battey. Oliver & Boyd, Edinburgh, UK, 323pp, 1972.*

Naturalist's guide to the Victoria region. *Jim West & David Stirling. Victoria Natural History Society, Victoria, BC, 200pp, 1986.*

Pacific northwest gold and gem atlas. *Bob & Cy Johnson. Private printing, Susanville, CA, 1985.*

Rock & mineral collecting in Canada — Yukon, British Columbia, Alberta, Saskatchewan, Manitoba. *Ann Sabina. Vol I, n1, GSC, Ottawa, ON, 1972.*

Rocks and minerals for the collector: the Alaska Highway; Dawson Creek, BC to Yukon/Alaska border. *Ann Sabina. GSC Paper 72-32, GSC, Ottawa, ON, 146pp, 1973, republished 1993.*

Rock and mineral collecting in British Columbia. *S.F. Leaming. GSC paper 72-53, GSC, Ottawa, ON, 138pp, 1973.*

Rockhounding & beachcombing on Vancouver Island. *Bill & Julie Hutchinson. The Rockhound Shop, Victoria, BC, 1971.*

Similkameen, the pictographic country. *N.L. Barlee. Canada West Publications, Summerland, BC, 96pp, 1978.*

Spud's dream: the story of how a Canadian mountain-man helped create a world-class city. *S.K. Cole. Rand & Sarah Publishing, Richmond, BC, 1987.*

Summary Report No.5: mica and vermiculite. *R.K. Collings, P.R. Andrews. CANMET, Ottawa, ON, 1991.*

The guide to gold panning in British Columbia. *2nd Ed., N.L. Barlee. Canada West Publications, Summerland, BC, 192pp, 1979.*

The geology of southern Vancouver Island: A field guide. *C.J. Yorath, H.W.Nasmith. Orca Book Publishers, Victoria, BC, 172pp, 1995.*

The identification of common rocks. *Information Circular 1987-5. E. Van der Flier-Keller, W.J. McMillan. Mineral Resources Division, Energy, Mines & Petroleum Resources, Victoria, BC, 17pp, 1987.*

The industrial potential of kyanite and garnet in British Columbia. *J. Pell. Province of British Columbia, Energy, Mines & Petroleum Resources, Victoria, BC, 1988.*

The rockhound's manual. *Gordon S. Fay. Harper & Row, NY, NY, 290pp, 1972.*

Treasure hunting in British Columbia. *Ron Purvis. McClelland & Stewart, Toronto, 1971.*

Vancouver Geology: a short guide. *G.H. Eisbacher. Geological Assoc. of Canada, Cordilleran Section, 56pp, 1973.*

Vancouver Geology. *John Armstrong. Geological Assoc. of Canada, Cordilleran Section. 128pp, 1990.*

Wagon road north. *Art Downs. Foremost Publishing, Surrey, BC, 80pp, 1969.*

Wet Coast Adventures: mine-finding on Vancouver Island. *Walter Guppy. Cappis Press, Victoria, BC. 192pp, 1988.*

West Coast Fossils. *R. Ludvigsen, G. Beard. Whitecap Books, Vancouver, BC, 195pp, 1994.*

Where terranes collide. *C.J. Yorath. Orca Book Publishers, Victoria, BC, 234pp, 1990.*

Yoho National Park, British Columbia: the mountains, the rocks, the scenery. *D.M. Baird. Misc. GSC Report No.4, 107pp, 1962.*

Publications of the 24th International Geological Congress, Montreal, 1972:
(available from the GSC, Ottawa)
>   Mineral Deposits along the Pacific Coast of Canada, *AC-06*
>   Copper & molybdenum deposits of the Western Cordillera, *AC-09*
>   Major lead-zinc deposits of Western Canada, *AC-24*
>   Coal, oil & gas and industrial mineral deposits of the Interior Plains, Foothills and Rocky Mountains of Alberta and British Columbia, *AC-25*

INDEX

Want to know more about these mineral localities?

This Field Guide is limited for space, so only a finite amount of information can be included with each site. If you have a DOS computer, and would like to know more about these sites, you should order the MINFILE/ pc program and the VANCOUVER ISLAND SPECIAL EDITION disk.

1. MINFILE/pc is a DOS database management system that allows you to search sites by geological, locational and economic variables. It comes with a substantial User Manual that will get you up and running in half an hour.

Order the MINFILE/pc operating system (the data management program) from MINFILE, Geological Survey Branch, Dept. Employment & Investment, 5th Floor, 1810 Blanshard Street, Victoria, BC V8V 1X4. Phone:(250) 952-0386, or fax: (250) 952-0381, or e-mail at ljones@galaxy.gov.bc.ca. You can also visit their webpage at http://www.ei.gov.bc.ca/geosmin/ minfile/minfile.htm and download the Vancouver Island Special Edition.

At the time of going to press, there was no charge for the product, and the package includes the MINFILE/pc disks (operates under DOS or Windows) and a manual.

2. The VANCOUVER ISLAND SPECIAL EDITION includes detailed geological reports, mine information, assay, reserve & production data, mineralogy and alternation descriptions, and literature references and publications.

Order the VANCOUVER ISLAND SPECIAL EDITION disk from Crown Publications Inc., 521 Fort Street, Victoria, BC V8W 1E7. Phone:(250) 386-4636, or fax (250) 386-0221, or reach them on the Internet at http://vvv.com/crownpub/

At the time of going to press, there was a charge of $7.50 plus GST and P&P.